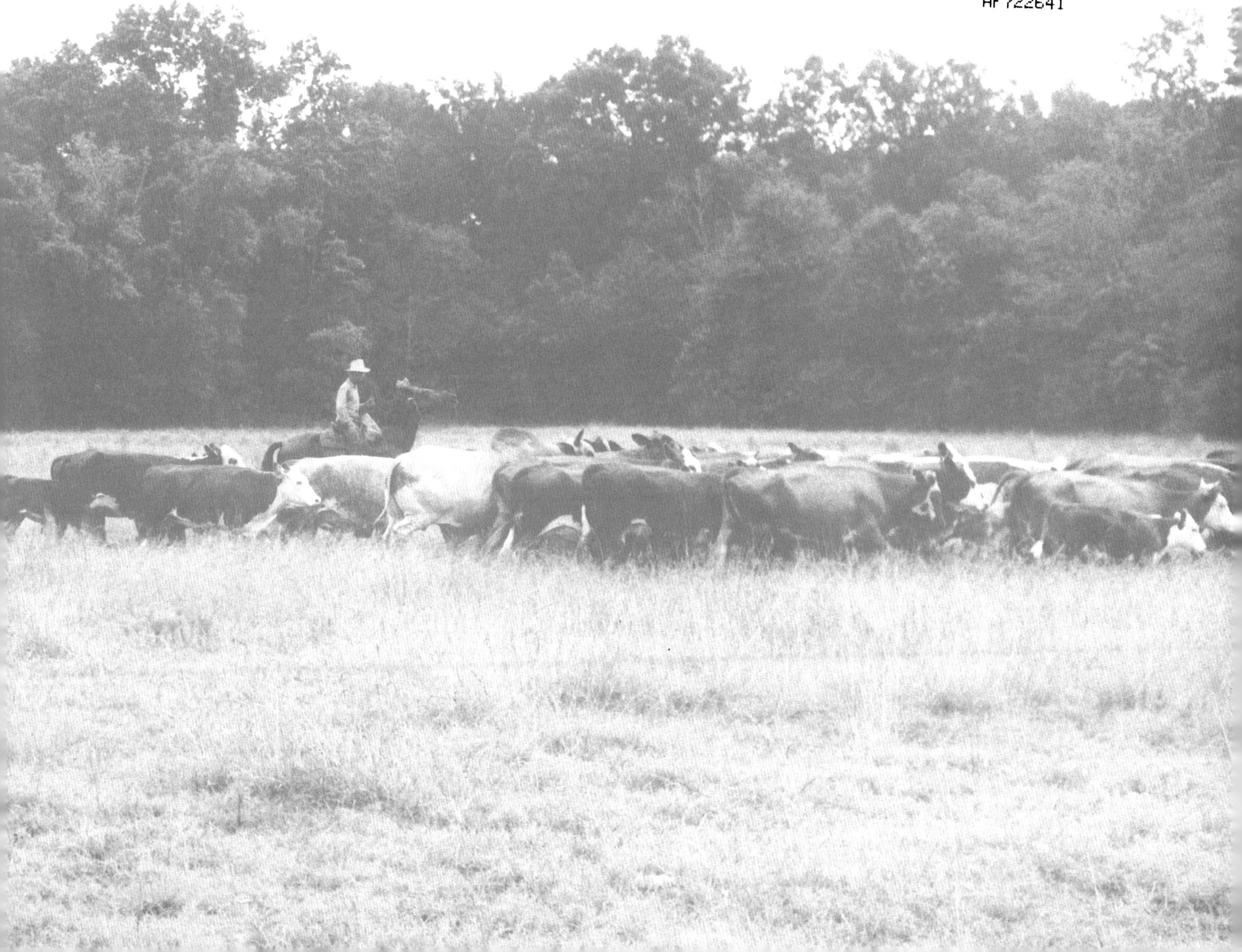

A Family of the Land

THE CHARLES M. RUSSELL CENTER SERIES
ON ART AND PHOTOGRAPHY OF THE AMERICAN WEST

B. BYRON PRICE, *General Editor*

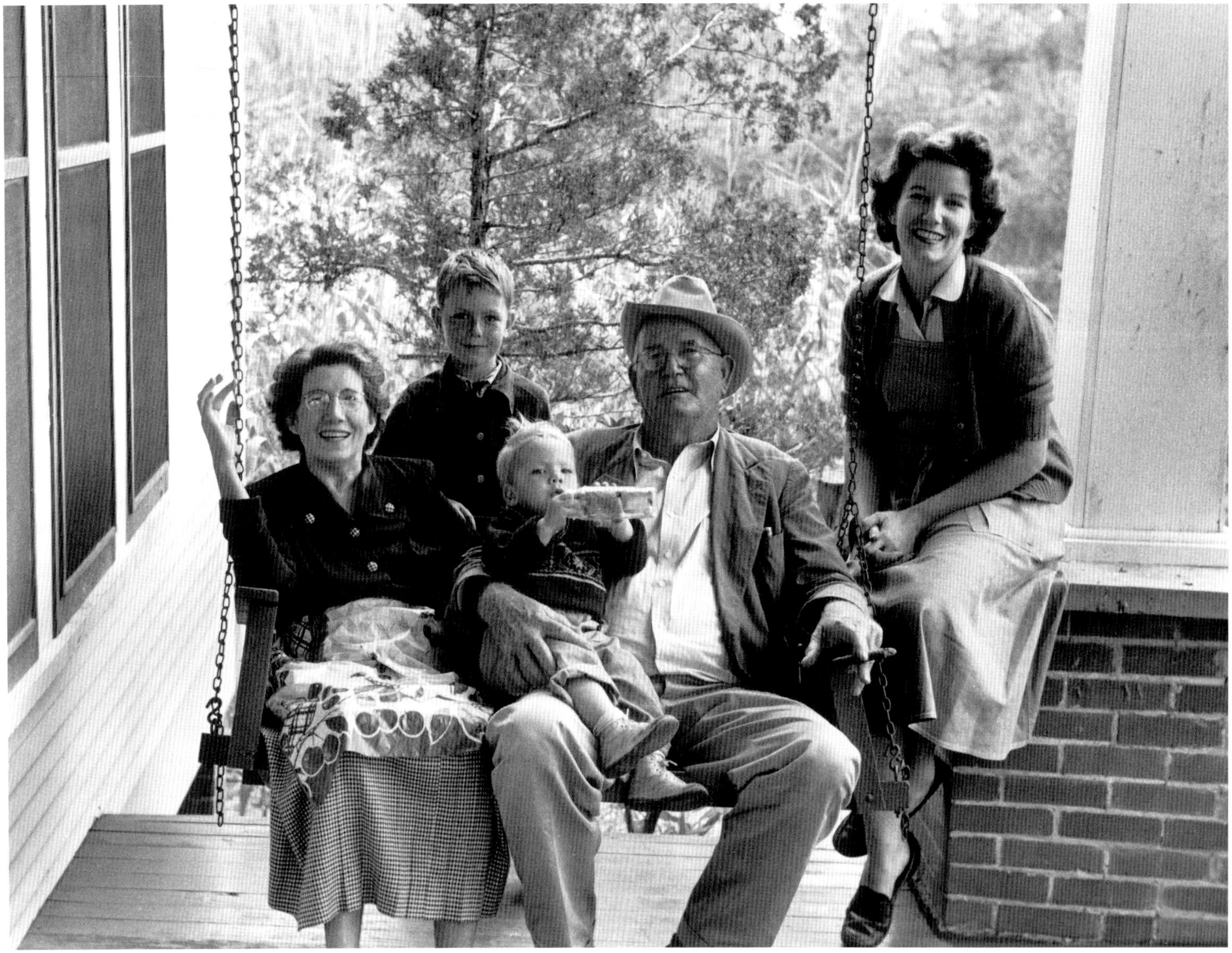

A Family of the Land

THE TEXAS PHOTOGRAPHY OF GUY GILLETTE

by ANDY WILKINSON

Foreword by B. BYRON PRICE

UNIVERSITY OF OKLAHOMA PRESS
NORMAN

A Family of the Land: The Texas Photography of Guy Gillette is Volume 13 in the Charles M. Russell Center Series on Art and Photography of the American West.

The paper in this book meets the guidelines for permanence and durability of the Committee on Production Guidelines for Book Longevity of the Council on Library Resources, Inc. ∞

1 2 3 4 5 6 7 8 9 10

LIBRARY OF CONGRESS CATALOGING-IN-PUBLICATION DATA

Wilkinson, Andy.
A family of the land: the Texas photography of Guy Gillette / by Andy Wilkinson; foreword by B. Byron Price.
pages cm.

(The Charles M. Russell Center series on art and photography of the American West; volume 13)

ISBN 978-0-8061-4404-7 (hardcover: alk. paper)

1. Crockett (Tex.)—History—20th century—Pictorial works. 2. Crockett Region (Tex.)—History, Local—Pictorial works. 3. Crockett (Tex.)—Social life and customs—20th century—Pictorial works. 4. Community life—Texas—Crockett—History—20th century—Pictorial works. 5. Ranch life—Texas—Crockett—History—20th century—Pictorial works. 6. Gillette, Guy—Family—Pictorial works. 7. Porter family—Pictorial works. 8. Ranchers—Texas—Crockett—Biography—Pictorial works. 9. Crockett (Tex.)—Biography—Pictorial works. 10. Gillette, Guy. I. Gillette, Guy, photographer. II. Title.

F394.C817W45 2013
976.4'235–dc23

2013009229

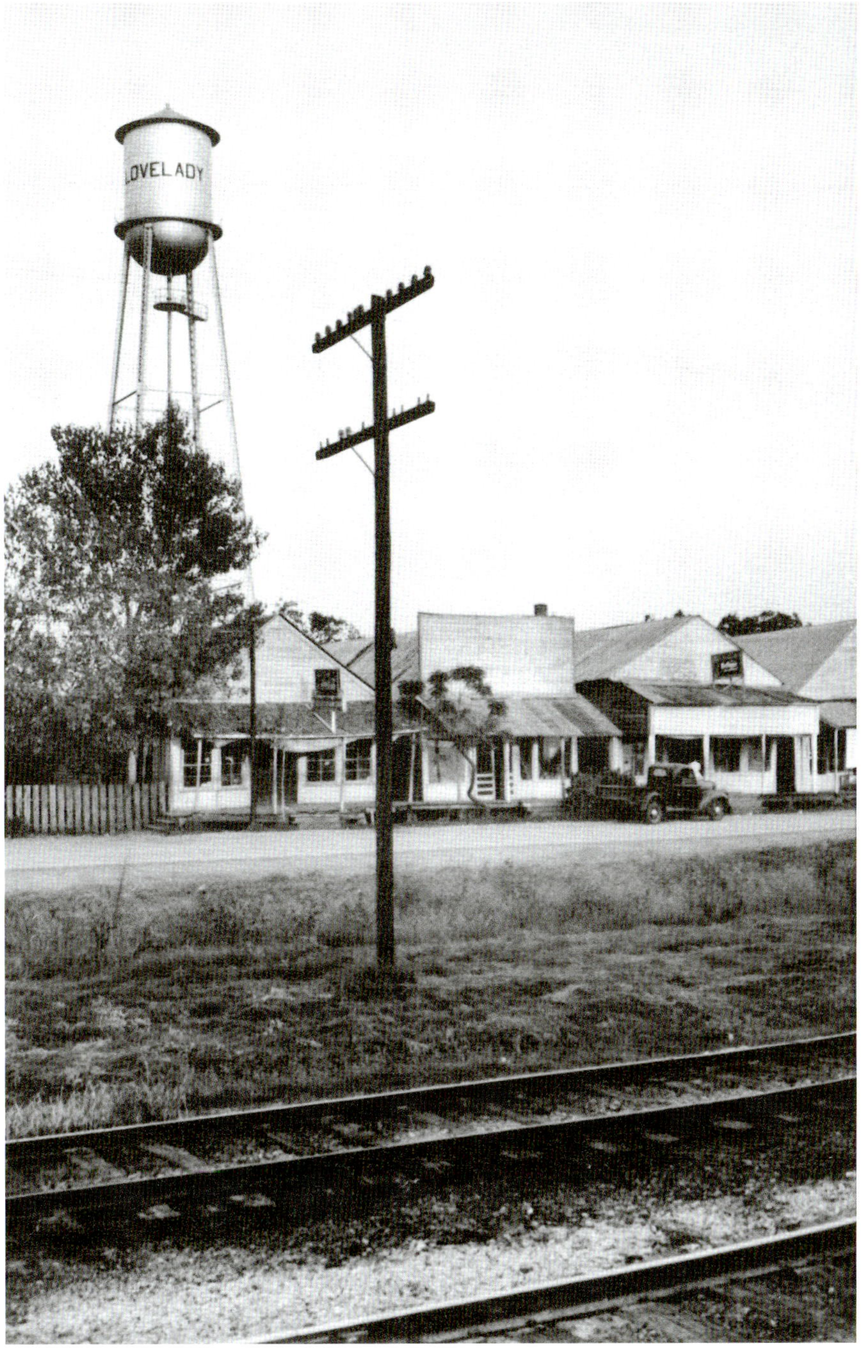

For Dorcie

GUY GILLETTE

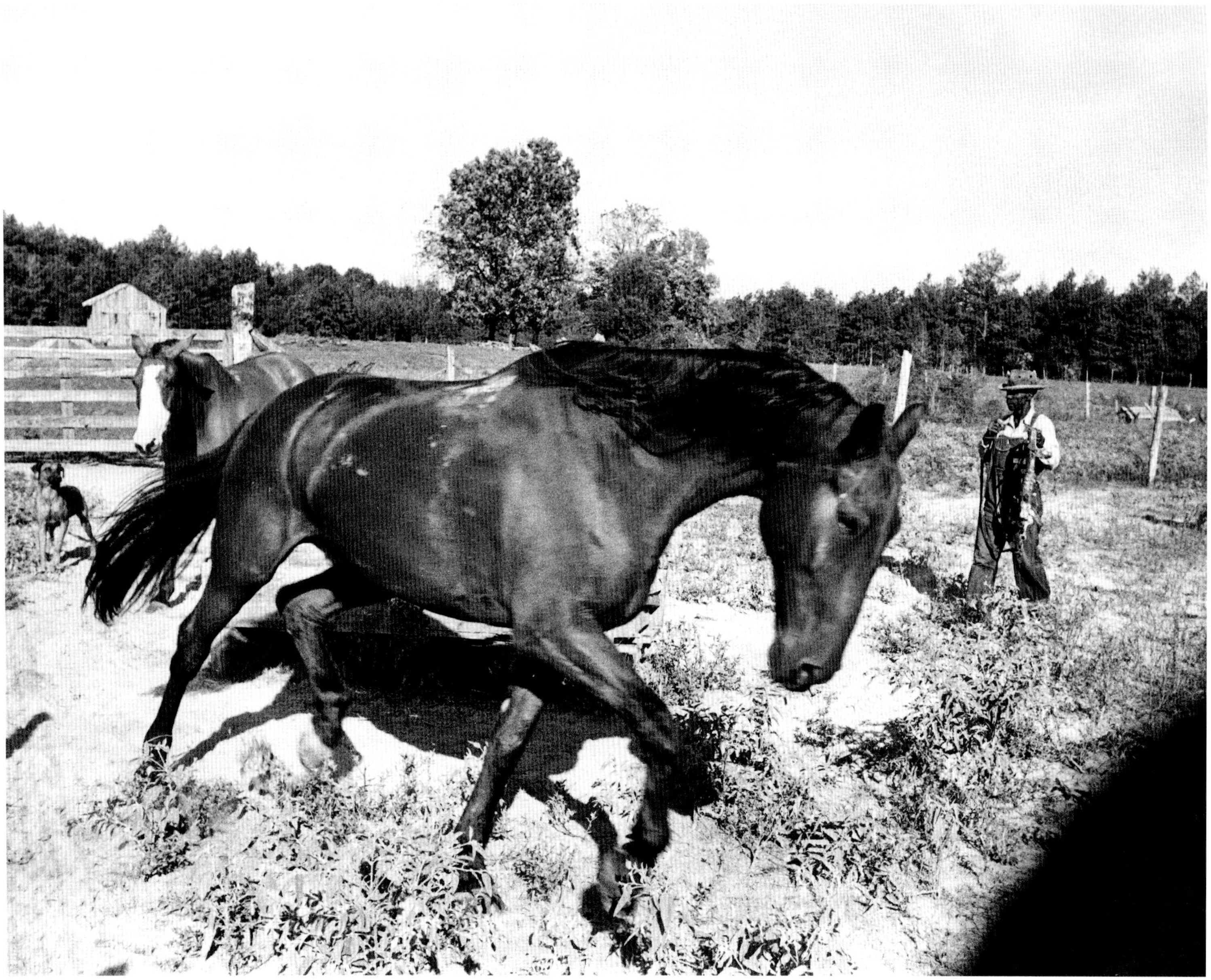

Contents

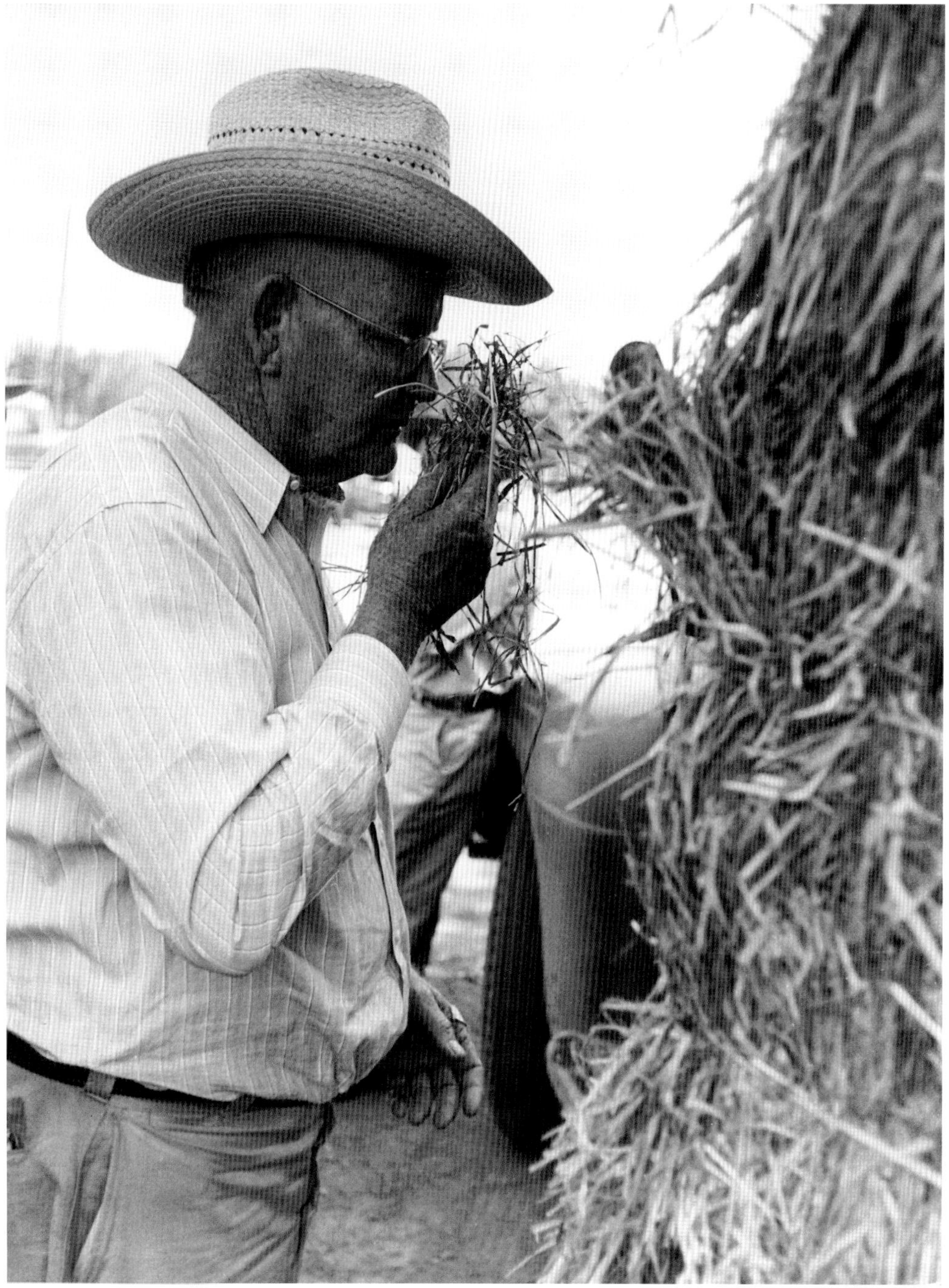

FOREWORD

STORYTELLING IS AT THE HEART OF THIS BOOK. With image and text as warp and weft and time and place as thread, photographer, Guy Gillette, and writer, Andy Wilkinson, weave an engaging multigenerational tale of family and community in the pine barrens of East Texas. The visual and written narrative that unfolds in these pages is deceptively simple. It flows across the landscape like a lazy river, smooth and unhurried, its irresistible power masked by the sights, sounds, and smells that mark its course.

Firmly rooted in the journalistic style reminiscent of the Farm Security Administration photographers of the Great Depression, Guy Gillette's imagery is decidedly more hopeful, offering glimpses of ordinary folk at work and play. His photographs introduce out-of-the-way towns with names like Crockett and Lovelady and gauge the rhythms of life on a gritty, sweat-stained cattle ranch, called the Porter Place. They are plain, straightforward images, much like the people who inhabit them, and although the rural life they reflect is neither quaint nor romantic, love and tradition abide there in equal measure.

Andy Wilkinson, a keen-eared singer, songwriter, and poet from the plains of West Texas, brings context to Gillette's photos of kith and kin. Like any skilled interpreter, Wilkinson appreciates the nuance and texture of language, even one that communicates visually rather than orally. His perceptive observations, reflections, and imaginings translate Gillette's work anew, and tease fresh meaning from the light and shadow of the past.

During the mid-1950s, the same era that Guy Gillette was snapping many of the images of family and friends in this volume, Swiss-born Robert Frank was crisscrossing the United States on a Guggenheim grant, photographing American life. His widely celebrated 1958 work, *The Americans*, portrayed a bleak and lonely nation at odds with itself and with most of the images in this book. Looking to reconcile these two disparate views, I was reminded of my old friend, the Texas novelist Elmer Kelton, who once observed that (and I am paraphrasing) upon entering a corral some people just naturally have their eye on the horses and others on the manure.

B. BYRON PRICE

A Family of the Land

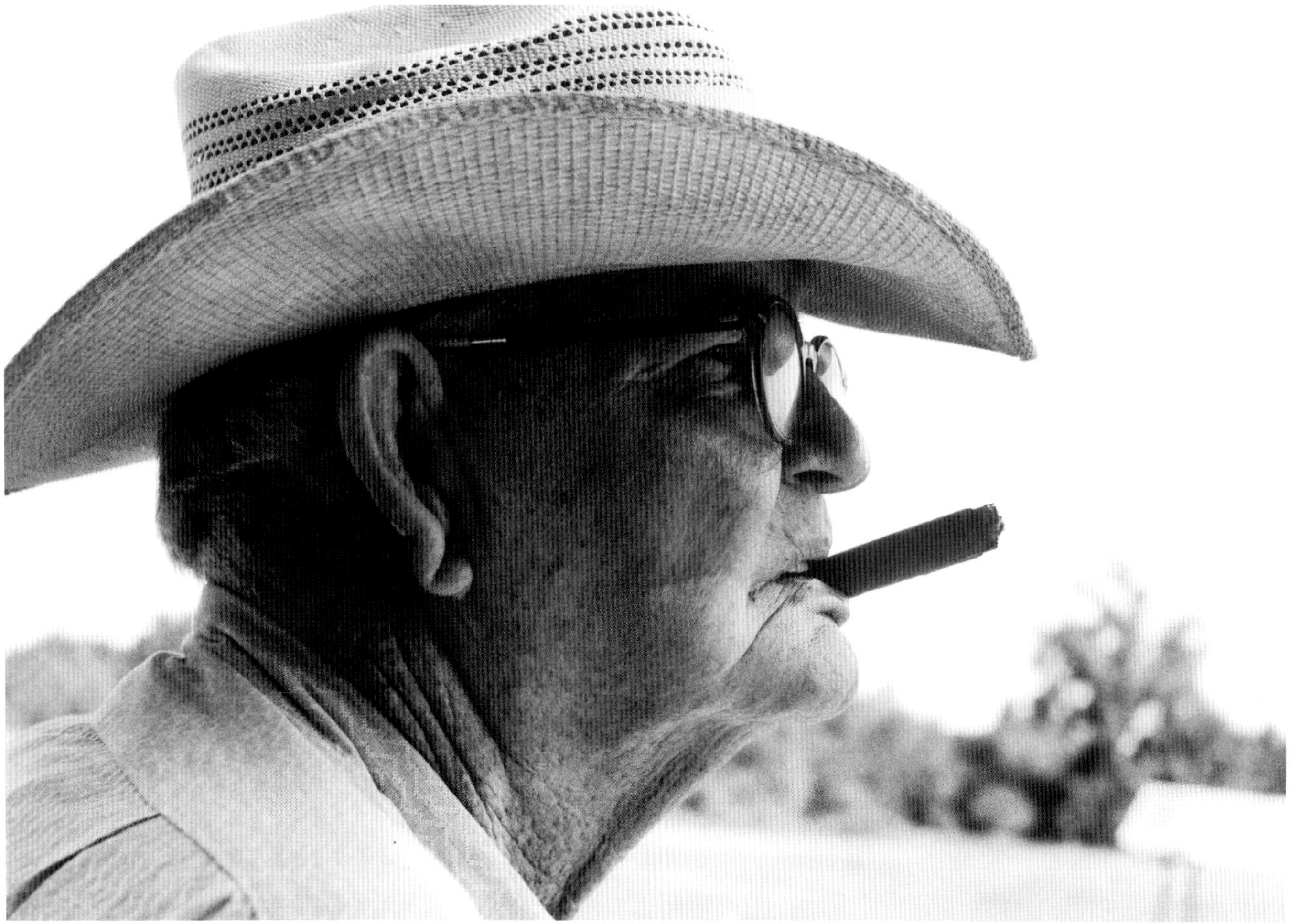

INTRODUCTION

In the deep Piney Woods of southeast Texas in the year before the publication of Cather's saga, young Hoyt Porter began assembling a ranch between the small town of Crockett and the smaller village of Lovelady. It's unlikely cowboy country, flanked by two lazy rivers — the Trinity on the west, the Neches on the east — and, with over forty inches average annual rainfall, a land requiring constant attention to fight back a relentlessly encroaching woodland.

Hoyt built his ranch in an unlikely way, too, for the likely narrative of American ranching history begins with wild and wide-open spaces that gradually lose both wilderness and openness as the juggernaut of settlement moves ever westward until only a few specimens of the real thing remain, hemmed-in by farms and towns and cities, criss-crossed by roads and jet contrails. But the Porter Place runs counter to that, cobbled together instead in the backwash of the westward movement, a quilt of square pastures stitched from the remnants of antebellum cotton plantations and sharecropper farms. Unlikely, but almost a century later it remains a working cattle outfit. And the Porter Place — Hoyt thought it pretentious to give his ranch a name — is still in the family, now run by the third generation to work it, Guy Porter Gillette and his brother, Pipp.

> *We come and go, but the land is always here, and the people who love it and understand it are the people who own it — for a little while.*
>
> WILLA CATHER
> *O Pioneers!* 1913

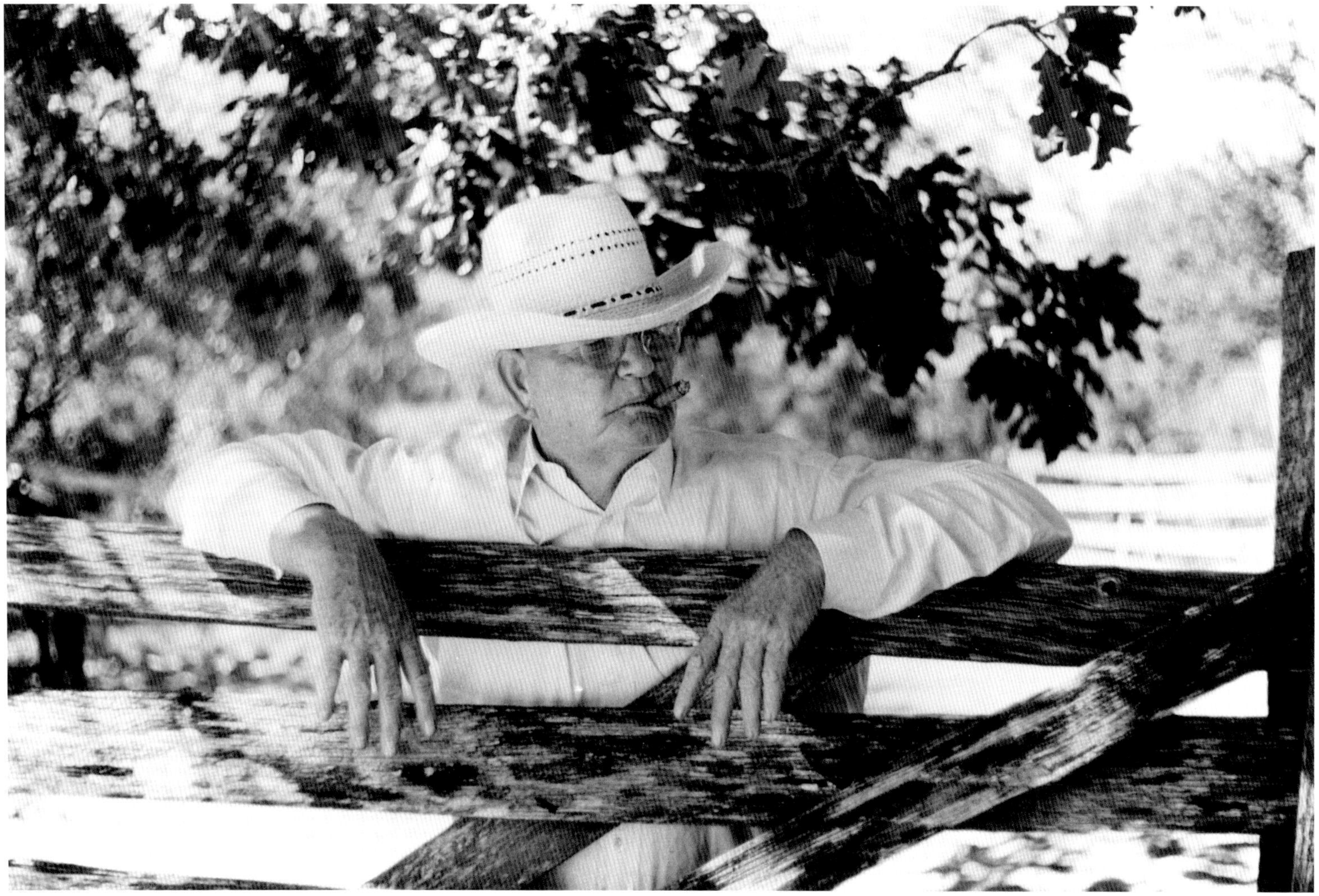

About the same time that he started acquiring land, Hoyt Porter married Lucy Thornton, a young girl from the nearby countryside. They had a daughter, Doris, in 1921, who grew up both on the ranch and in the mercantile store that her parents owned and operated in Lovelady. Doris read the magazines and studied the dress patterns that numbered among the store's goods and, having a talent for drawing, resolved to become a designer. She moved to New York City in 1941 to attend the prestigious Traphagen School of Design. To support herself, she took a job as a waitress in a vegetarian diner on West 56th, a place frequented by students from the Chekov Acting Studio. One of them, Guy Gillette, found work at the eatery as a busboy and soon found love there as well. Doris and Guy married in 1942 and made their first trip to Texas the following year. After their son Guy Porter was born in Crockett in 1945, they began a long series of summer trips between their home in New York and the Porter Place.

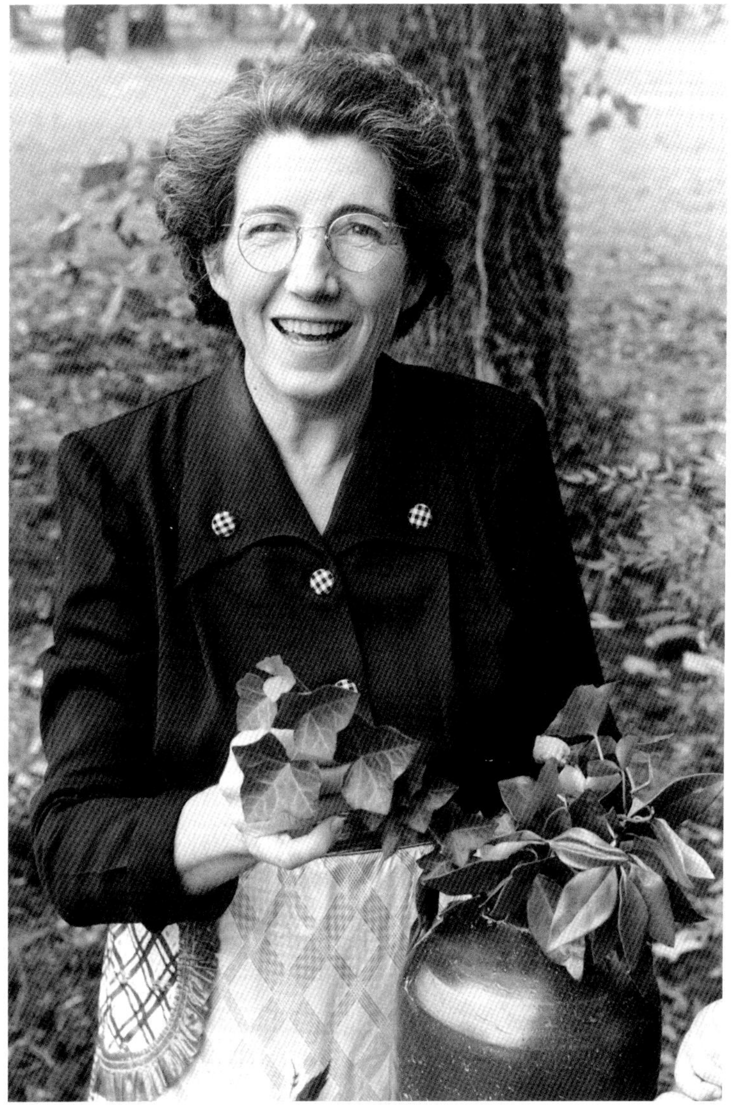

Besides his penchant for acting, Gillette, born in 1922 in Minnesota, also had an early interest in photography, so it seemed natural to document his family ranch trips. His eye was natural, too, and from the start he produced quietly powerful black-and-white images that told the story of his family and the land on which they lived, images framed by an actor's sense of movement and composition, images enriched by an artist's attention to detail and tonal value. A magazine editor offered him an assignment after having seen some of these Texas photographs. Guy had to borrow

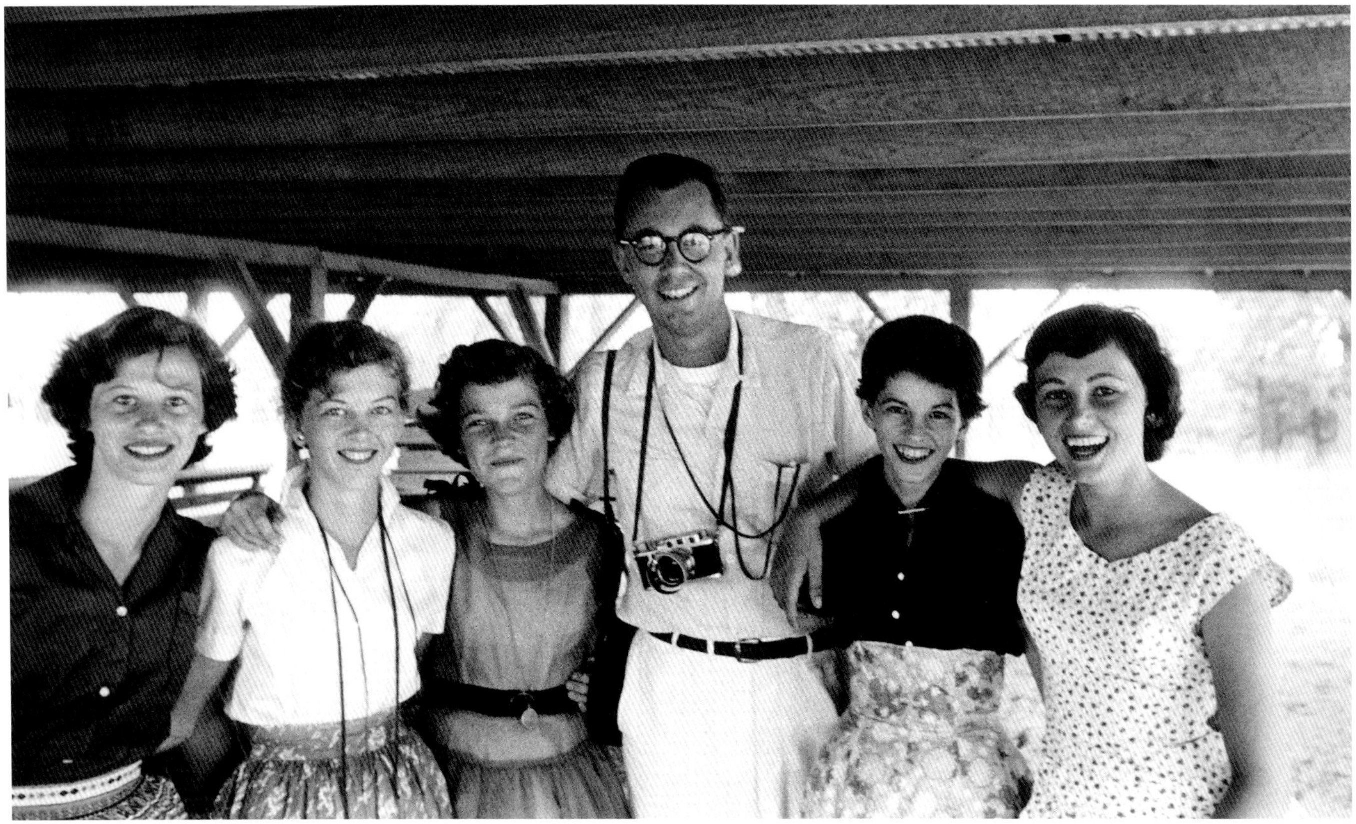

money to buy a good camera but was soon doing work for the best magazines of the day, such as *Fortune, Life, Harpers Bazaar, Theatre Arts*, and *This Week*. Seeing the possibilities for a career, he spent two years studying with Sid Grossman at the Photo League, and his photography began attracting attention, most noticeably from Edward Steichen, who featured two of Gillette's pieces in the *Family of Man* exhibit at New York's Museum of Modern Art (1955) and in the book that followed. One of those photographs was a family portrait made at the ranch.

Gillette's body of work remains highly regarded, with a recent exhibit at the Hudson River Museum, a full feature story in the March 2010 issue of *B&W Magazine*, and current representation by the Monroe Gallery of Photography in Santa Fe. His long career has covered a wide range of subjects, from front-line photojournalism — the Korean War, Civil Rights marches, Vietnam War protests — to sensitive and sophisticated portraits of the famous and the ordinary. He is particularly well-known for his images of actors, among them Charlie Chaplin, Audrey Hepburn, Lillian and Dorothy Gish, and Diane Keaton. But throughout, there have always run the entwined threads of family and land, most evident in several hundreds of images of the Porter Place.

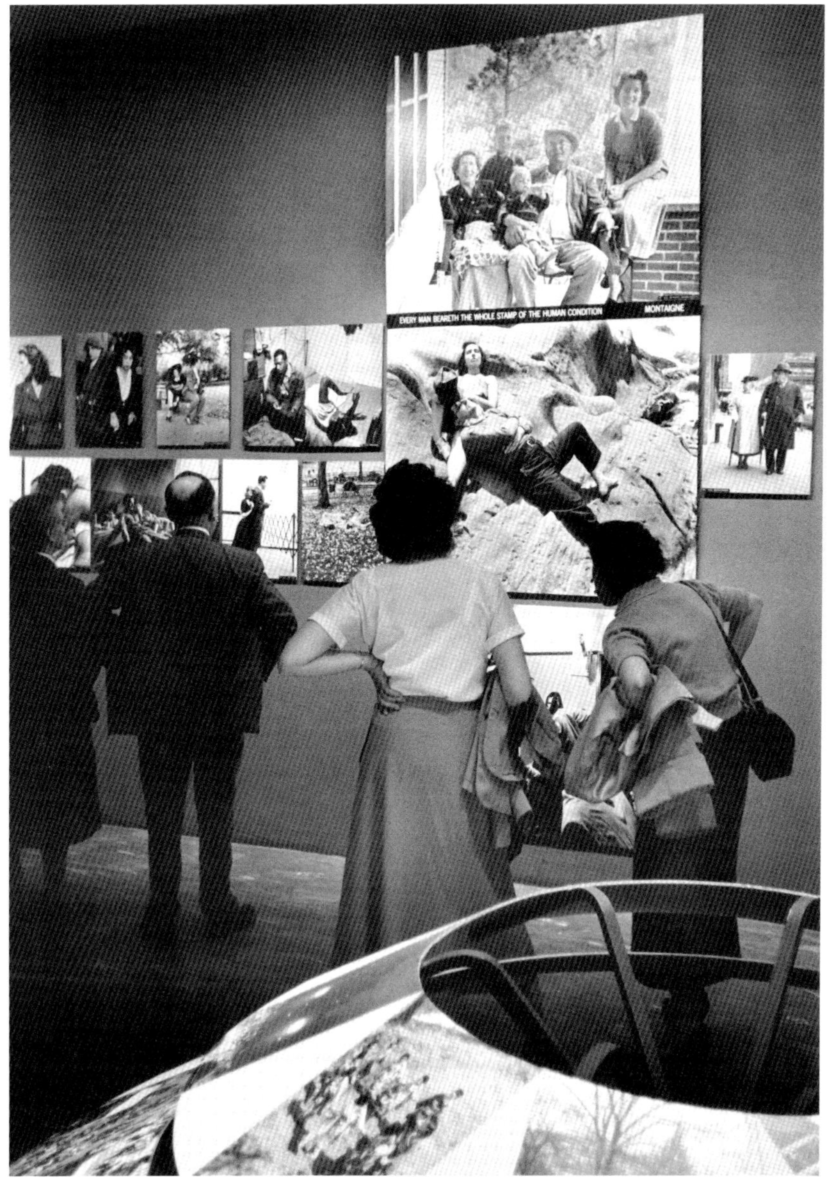

Lovelady, Texas, 1950s

There are things to talk about, squatting on the stoop above the sidewalk, leaned up against the doorway. Talk, because nothing else can be done. Weather first. Hot enough for ya? Weren't that a good rain? Think it's ever gonna rain again? Think it'll ever quit raining? Casual talk in the city, but not here. In Houston or San Antonio or Dallas, weather can wreck your plans for the day, for the weekend. An annoyance when it's not good, taken for granted when it is. Here it can wreck your season, your life. It can put money in your pocket, or break you.

Prices next. Can't do much about those, either, when the traders in the commodity exchanges are in Chicago and New York, the buyers in London and Los Angeles. Maybe it's not time to sell. Maybe it is. Maybe should've sold already. What's your neighbor doing? Weather and prices, talk that is acknowledgment, not supplication, less thanks or complaint than the admission that we all labor under a higher power. Like saying grace before supper.

And there is much to say grace over in Lovelady now. Cotton wasn't worth a dime in the 1930s, but it doubled during the war and is headed up into the low-thirty-cent range. Cattle, too, rose steadily in the 1940s then took a big jump when the troops started coming home. Situated some fifteen miles south of the county seat in Crockett and five miles north of the county line on the highway to Huntsville and with a stop on the Houston and Great Northern, the town's in a good location to recover from the Depression. Settled in 1872 when the railroad came through, by the turn of the century it was an important shipping point for cotton and corn and cattle. It had a school and a newspaper and several churches and hotels. By 1930, the population was over five hundred and there were just under fifty businesses. It's not that now, and yet . . .

The Magnolia Cafe is still serving the blue plate special. You can have a Pepsi-Cola with it, or an Alamo soda. Then you can walk

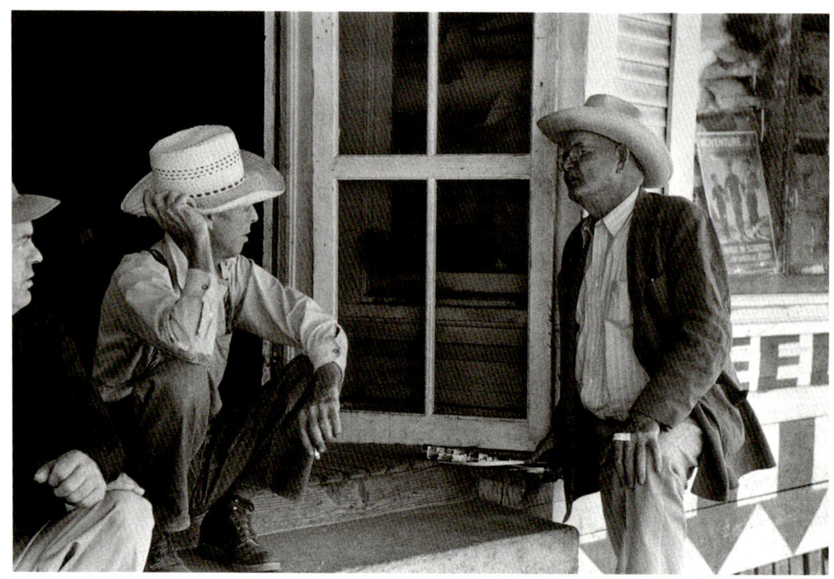

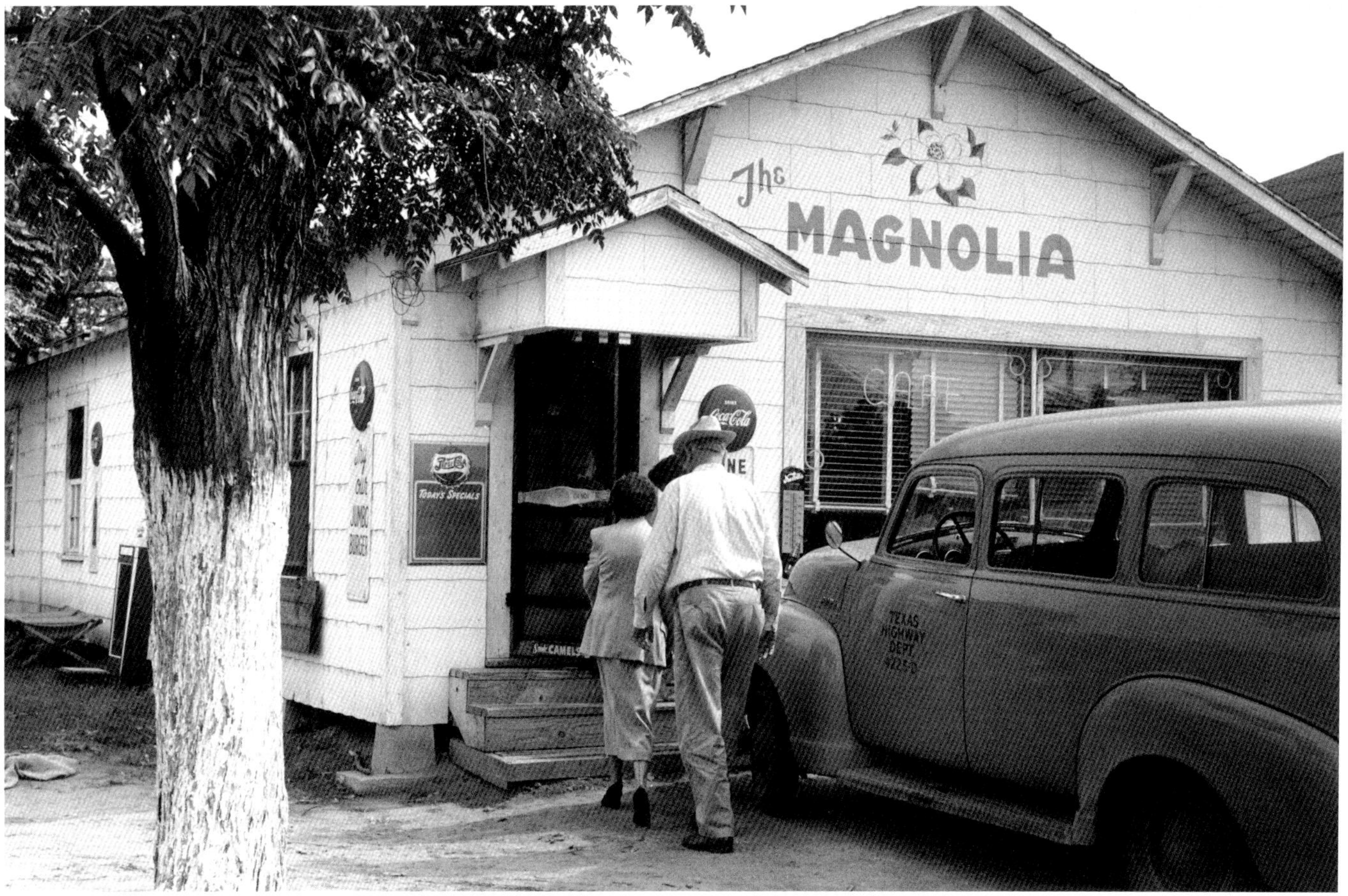
The
MAGNOLIA
Coca-Cola
TODAY'S SPECIALS
TEXAS
HIGHWAY
DEPT

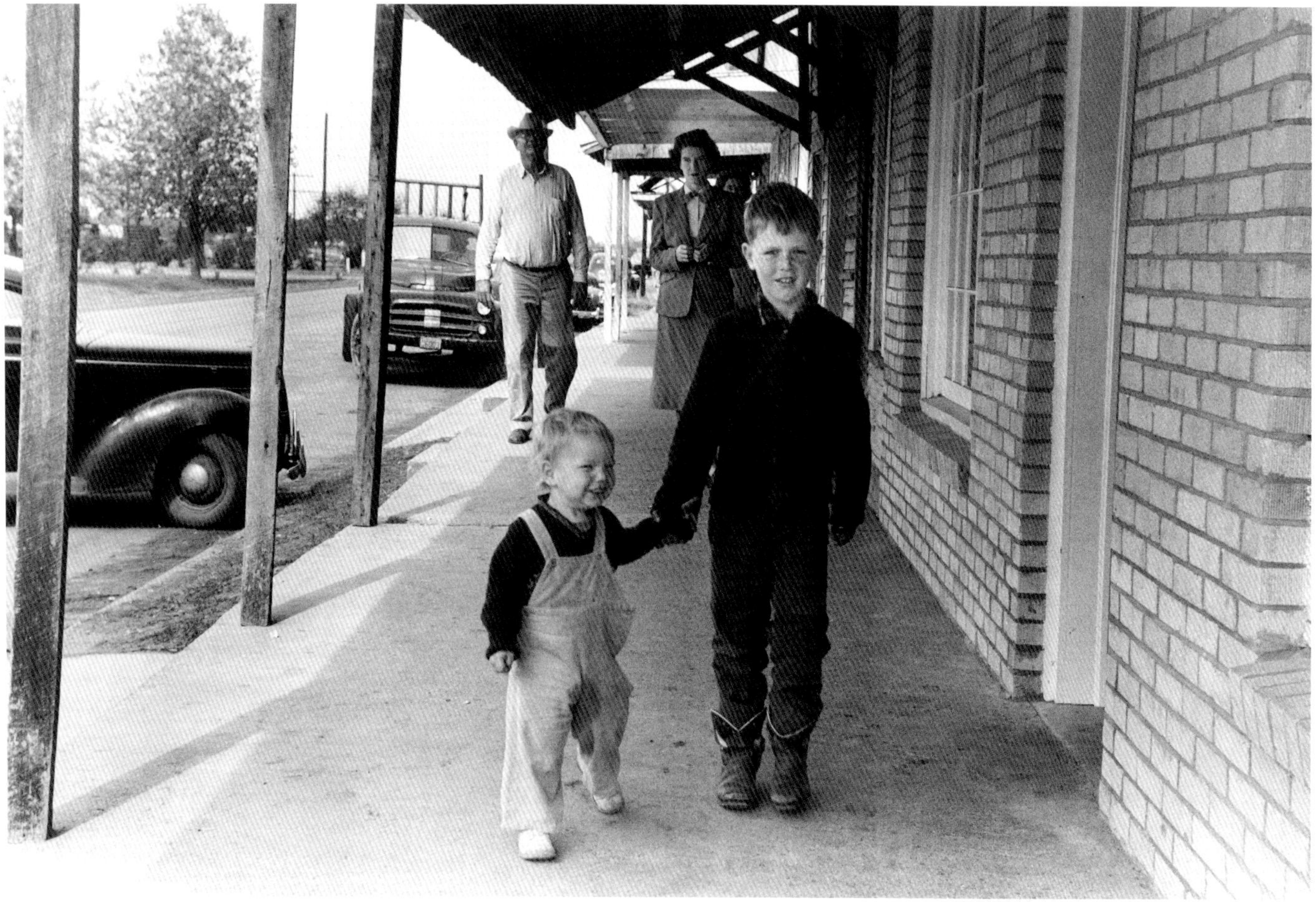

down the street to Porter's, where they're still selling groceries and dry goods and shoes. And magazines. Young Doris read them when she worked there for her folks, finding an advertisement in *Vogue* for a school of design in New York City. Now back with Guy and the boys for their annual summer visit, she's both beautiful and city-smart, well-put-together but comfortable alongside her dad's khakis and Stetson. Her level gaze puts one Hollywood myth to rest: small-town girls are born clear-eyed, not wide-eyed.

Arnold's Cafe is open, too, selling Coca-Cola — say that co-CO-la — a king-size one for six cents. The drug store's still in business. So's Parker's and King's and Stansel's. The high school, home of the Fighting Lions, is still big enough to field a marching band. There's no hotel, but the flag's still waving.

There is much to say grace over.

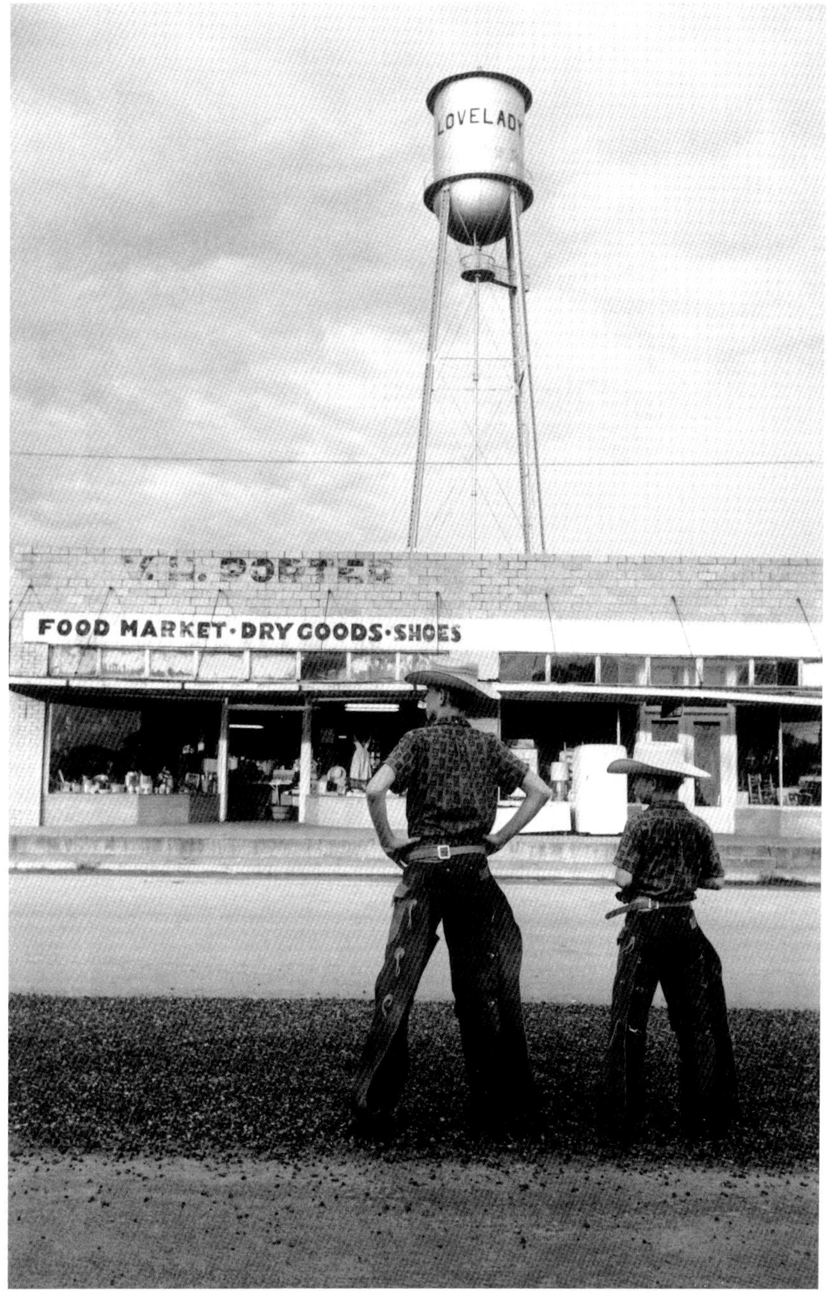

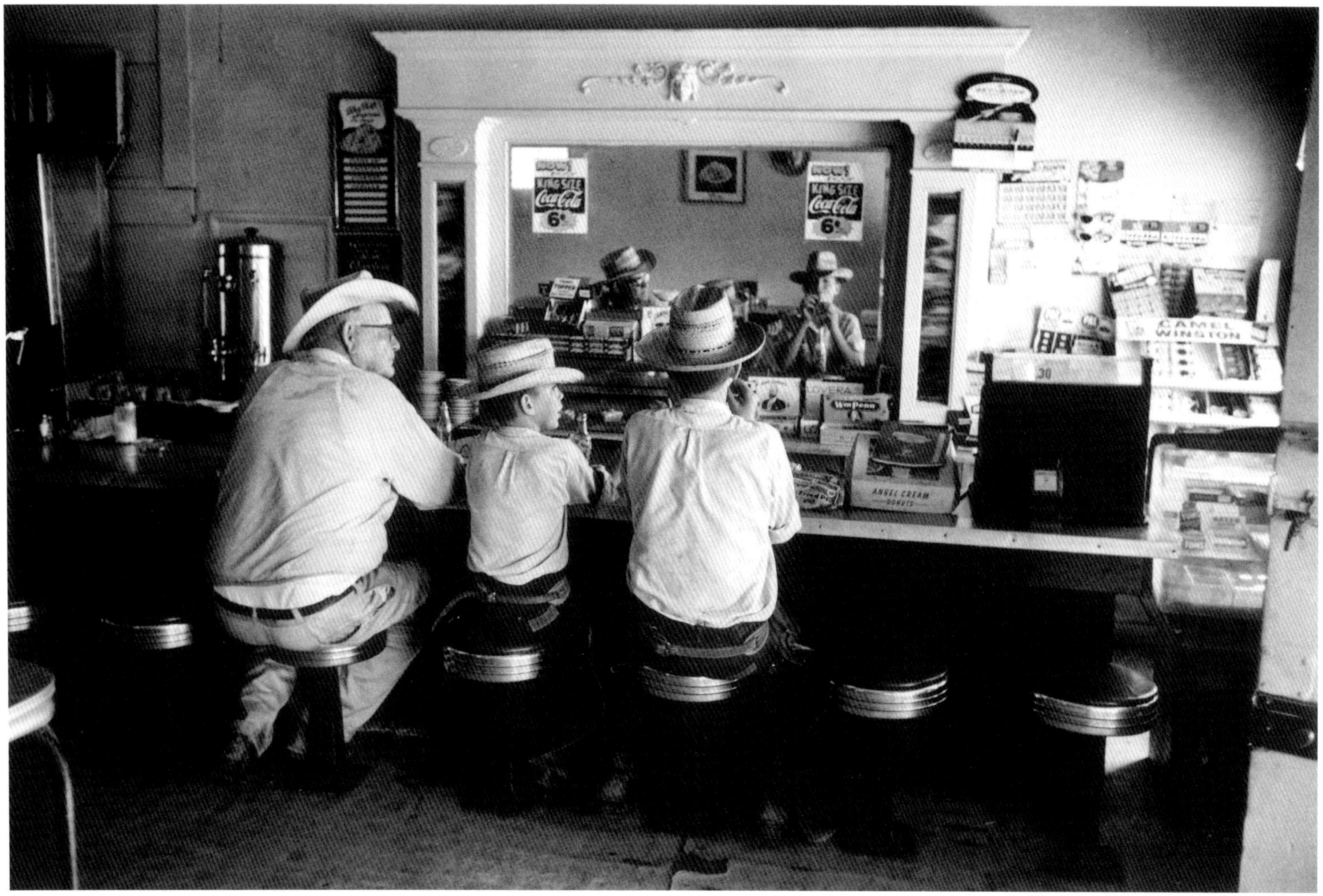
KING SIZE
Coca-Cola
6¢
KING SIZE
Coca-Cola
6¢
CAMEL
WINSTON
ANGEL CREAM
DONUTS

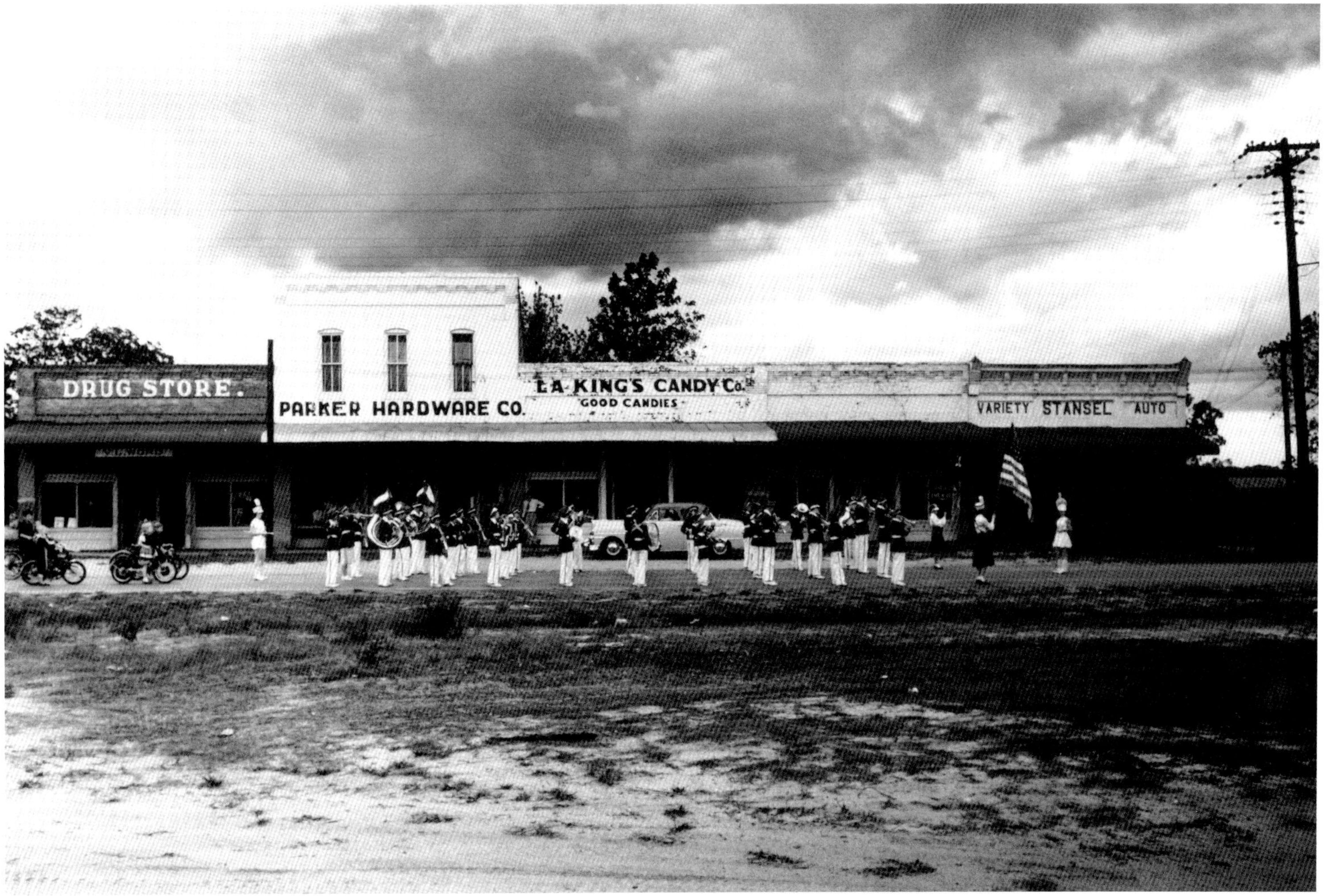
DRUG STORE.
PARKER HARDWARE CO.
E.A. KING'S CANDY Co.
GOOD CANDIES
VARIETY STANSEL AUTO

Saturday in Crockett

In the time before shopping malls and strip centers, before chain stores and convenience stores and box stores, people came to the county seat on Saturdays. From the smaller towns and communities, from the farms and ranches, they collected on the square, stores and cafes and pharmacies and barbershops lining the streets around the courthouse in the center, a distinctly American version of the traditional European plaza, a place where local government coexisted with local business, a place where commerce met fellowship. People dressed up. Not in Sunday-go-to-meetin' outfits, but not in work clothes either. Women in dresses and low-heeled, comfortable shoes, carrying purses and babies, the girls in jumpers or skirts with their hair pinned back or in scarves. Men in shirtsleeves but wearing clean, creased felt hats, the boys in freshly ironed shirts, faces washed and hair combed.

The square in Crockett vibrates with long and good use. Bounded by Houston Avenue on the north, Fourth Street on the west and Fifth Street on the east, on the south side is Goliad Avenue, part of El Camino Real, the first road in Texas, laid out by the Spanish in 1690. Crockett is the seat of Houston County, the first organized by the Republic of Texas, named for Sam Houston, the new republic's president and the hero who whipped Santa Ana near Lynchburg Ferry in the eighteen-minute-long Battle of San Jacinto. The town carries a hero's name, too. Davy Crockett camped at a spring there early in 1836 on his way to the Alamo.

A cotton plantation economy, Houston County grew quickly. At the beginning of the Civil War it numbered eight thousand, including almost three thousand slaves, making it one of the major population centers of the state. Though *center* isn't truly descriptive, as the great mass of its folk lived outside of the county seat, scattered over the county's twelve hundred square miles, mostly along the two rivers that mark its eastern and western lines and along its one road. After the war, Reconstruction replaced a morally bankrupt economy with an economy that was simply bankrupt. Times were thin for black and white alike, and Houston County did no more than hold its own until the railroad came through and jump-started the economy. Population doubled by 1880, almost doubled again by the turn of the century, peaking in 1930. Then a decade-long drought and the Great Depression set everything back until the prices of agricultural goods began to rise during the Second World War.

By the time that Guy Gillette came to Crockett, the square was back to its Saturday finest. He brought his camera and a way of seeing the life of the street and the city and the town. He'd been learning how to see in this way since he was a teenaged kid. The topic of his first photography project was Chicago, the city that works. And he was fresh from studying with Sid Grossman and the best of America's documentary photographers at the Photo League in New York. By this particular Saturday, Guy and the square were at work.

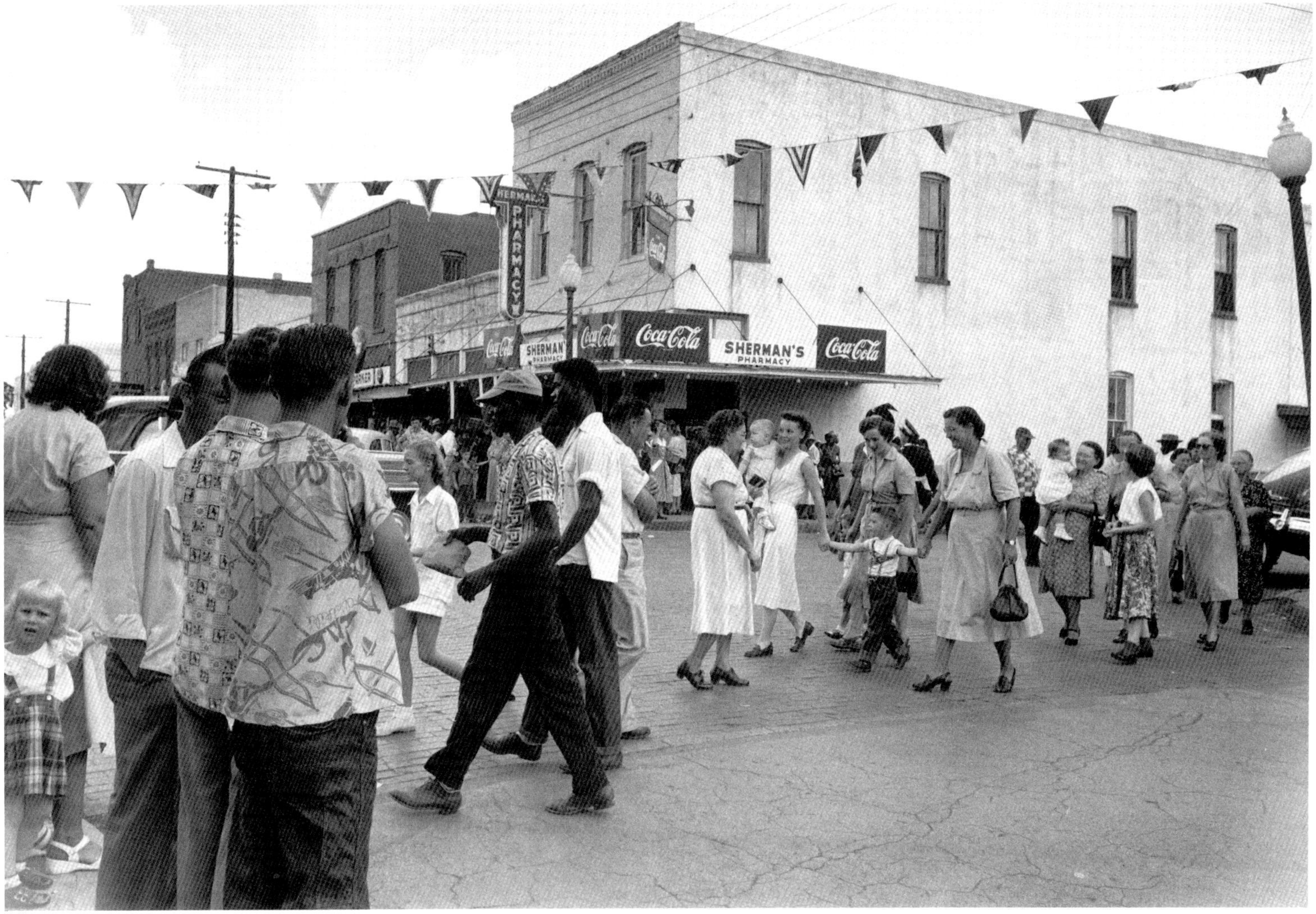
PHARMACY
Coca-Cola
SHERMAN'S PHARMACY
Coca-Cola

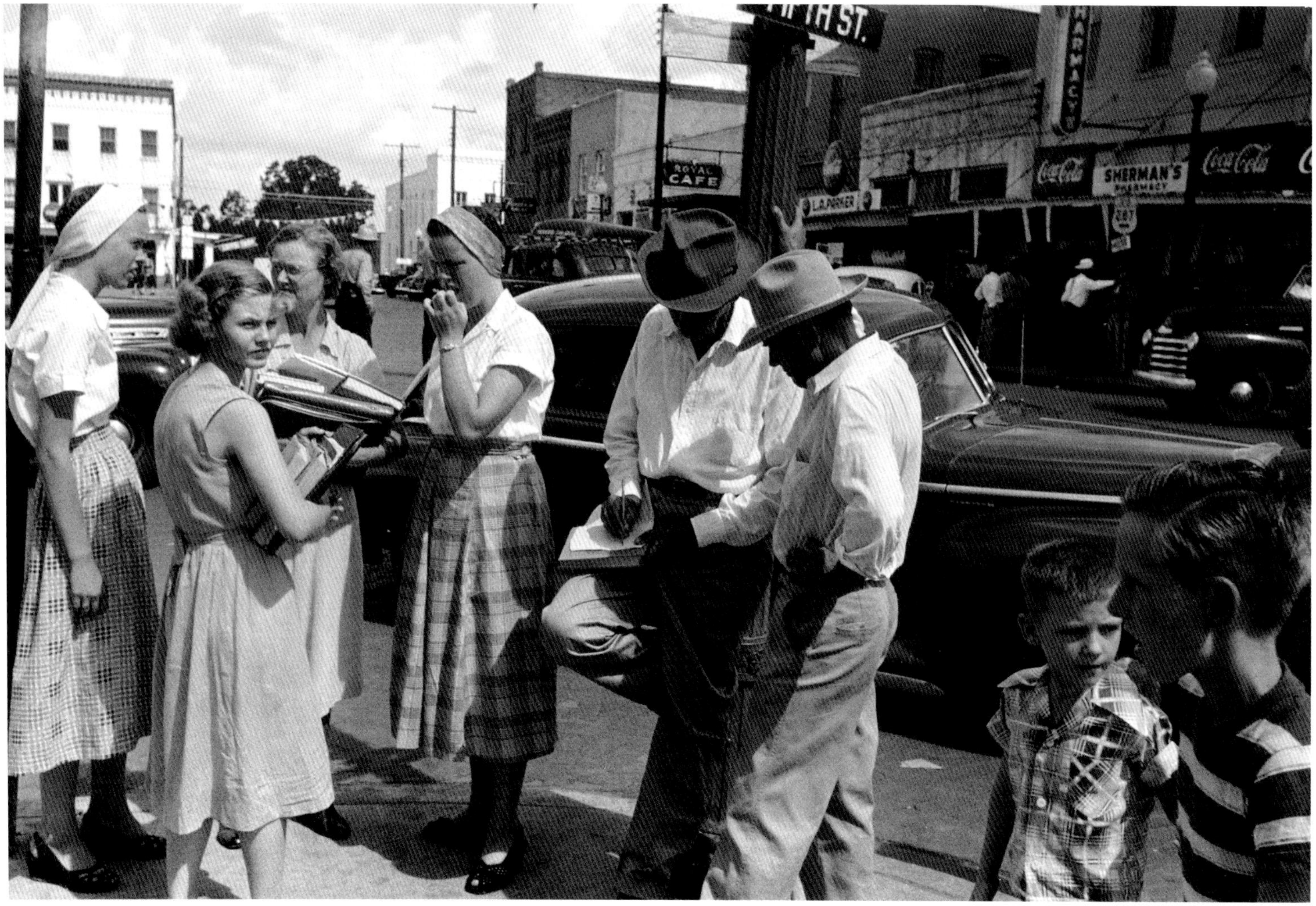
ROYAL
CAFE
Coca-Cola
SHERMAN'S
Coca-Cola

Try as I have — and I have tried several times — to come up with a pithy, literary description of what photography means to me, the years have come and gone and I still cannot quite manage it. So I will rely on the words of others. Jacob Deschin, writer and photography critic for the *New York Times*, said that the true photographer is alert, tries to understand what is going on, then readies himself to record the meaningful moments.

Sid Grossman, who taught at the Photo League in New York, stressed that the viewfinder is your world. You, the photographer, are responsible for all that appears in your photograph. It is your viewfinder, a whole world that is a jumble unless you bring order into it, an order that is always significant to you.

And I hold with the French photographer Brassaï. I do not look for exceptional subjects. I avoid them. I think it is daily life that is the great event, the true reality.

Striving for the meaningful moment, the great event, these things have occupied me fully my whole life. It is why I have enjoyed watching the people of Houston County, seeing them through a camera's viewfinder.

GUY GILLETTE

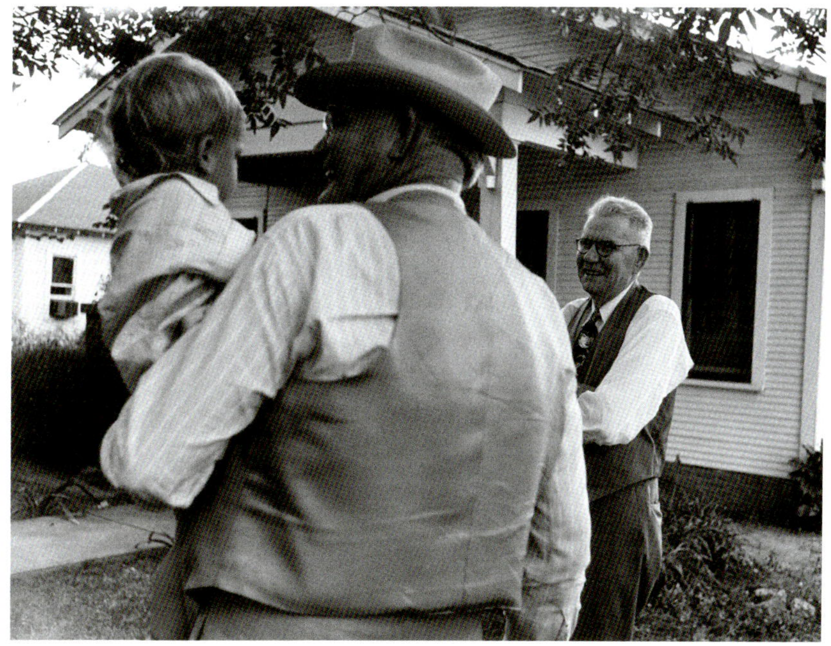

Hoyt and Claude

The land has its own tempo, its own pace, an unfolding of time that owes nothing to the manager's schedule. People of the land — those who love it and understand it — let their works and days move with its rhythm, an ebb and flow with never a boundary set or regulated by clock and calendar, by memorandum and committee meeting. For the people of the land the day begins with sunrise, not with the alarm bell. It ends with sunset, not with the afternoon rush hour. Dinner is at noon, when the sun, not the hour hand, stands straight up. Supper is at twilight.

Hoyt and his friend Claude Lundy have business in Centerville, thirty-three miles west of Crockett, an hour's drive by pickup truck from Lovelady. A parcel of land to look at, to consider. Or a bull for sale, or some yearlings. The pickup's 2-50 air-conditioning is on full blast. That's two windows down, fifty miles an hour. It's the coolest they'll be all day.

They have time to stop for dinner. Their hats come off when the food's set down. Today it's a plate lunch, with store-bought white bread stacked on a dish next to the bottle of pepper sauce. There's coffee and cobbler. Coffee, always, even in hot weather. Hoyt saucers and blows his. The noon sun fills the cafe, all mid-tones and back-lighting, turning the formica table tops into mirrors, a tough challenge for the photographer.

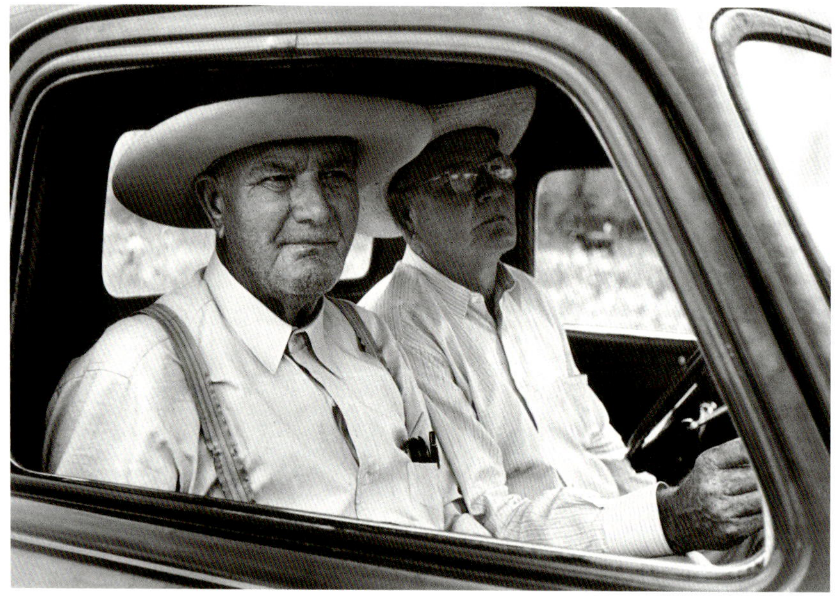

Leaving the cafe, Hoyt and Claude have time to stop for a visit by the bench outside Johnson's Grocery & Market. That parcel of land isn't going anywhere soon. Neither is the bull, or the yearlings. There's plenty of time.

Work done to the rhythm of the land is productive. The sine waves of the days weave into the seasons, the seasons into the years, the years into the big circles of life. These waves, this rhythm of the land, form the arc of the narrative, the one true story, told and retold, over and over and over. Story is process, not product, but trust in the process and the product takes care of itself. The rhythm of the land is productive.

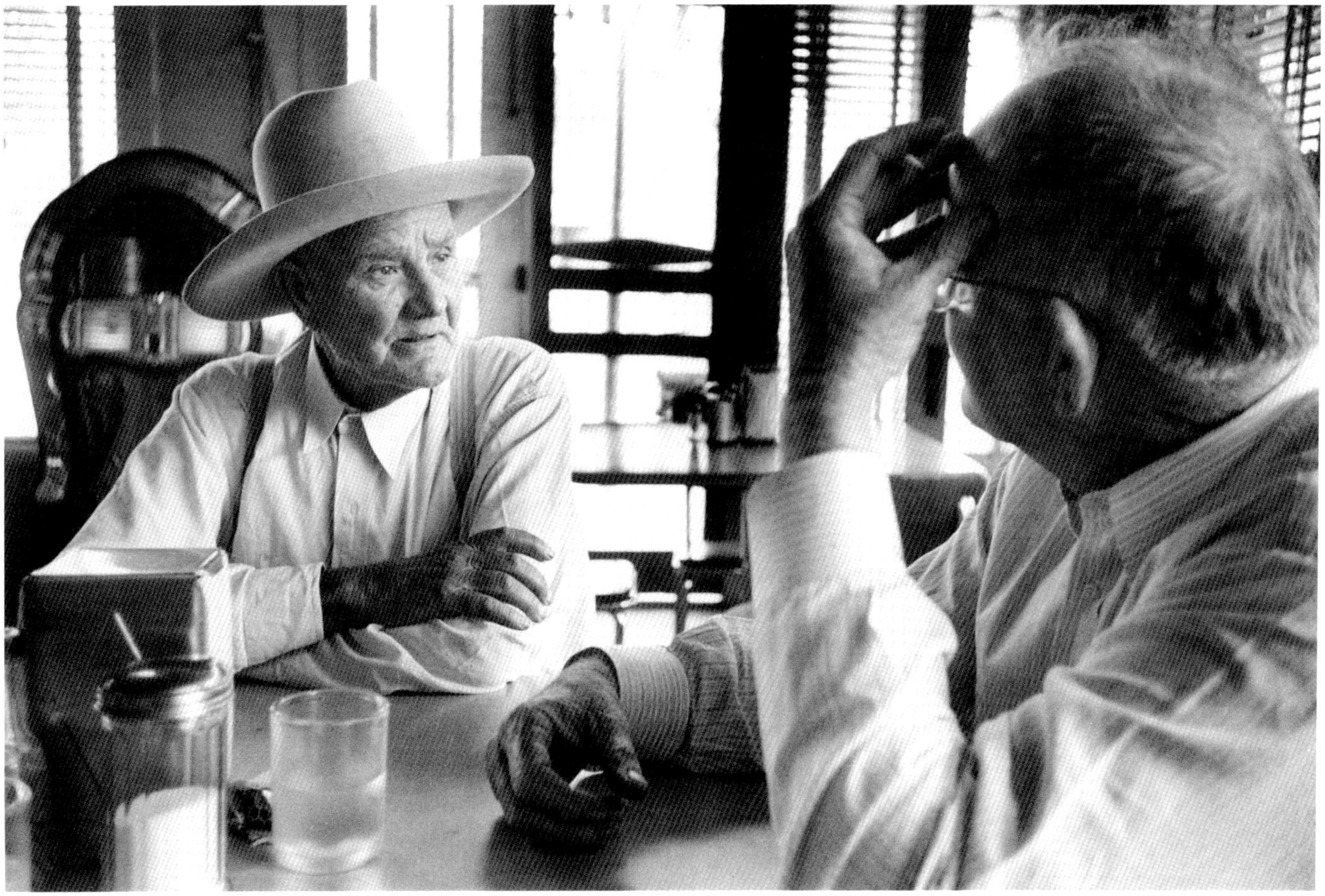

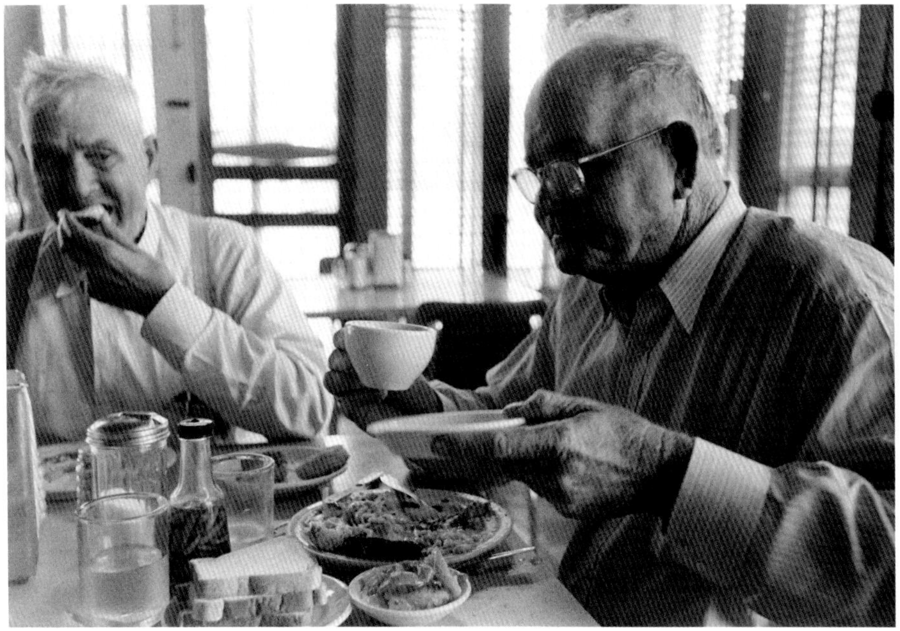

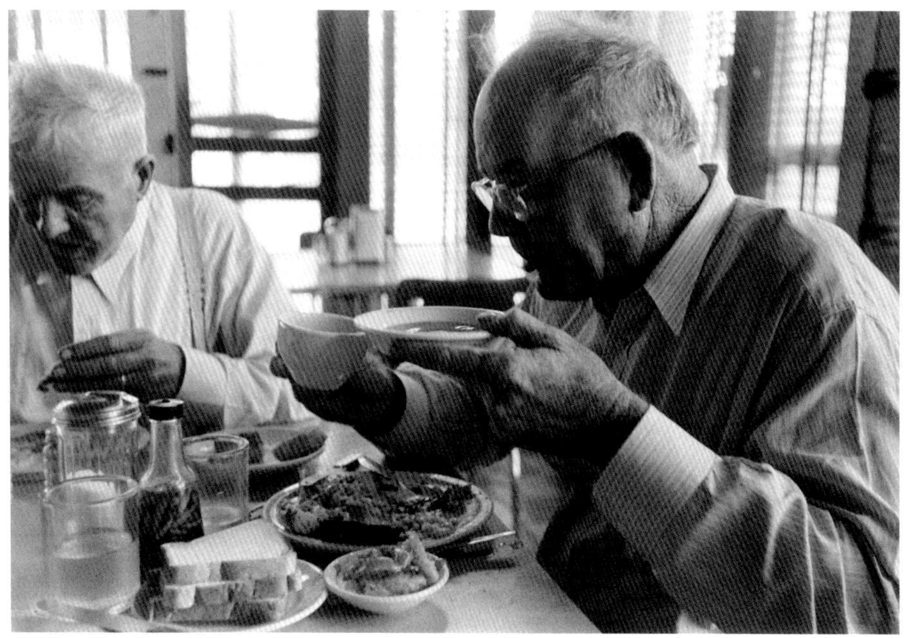

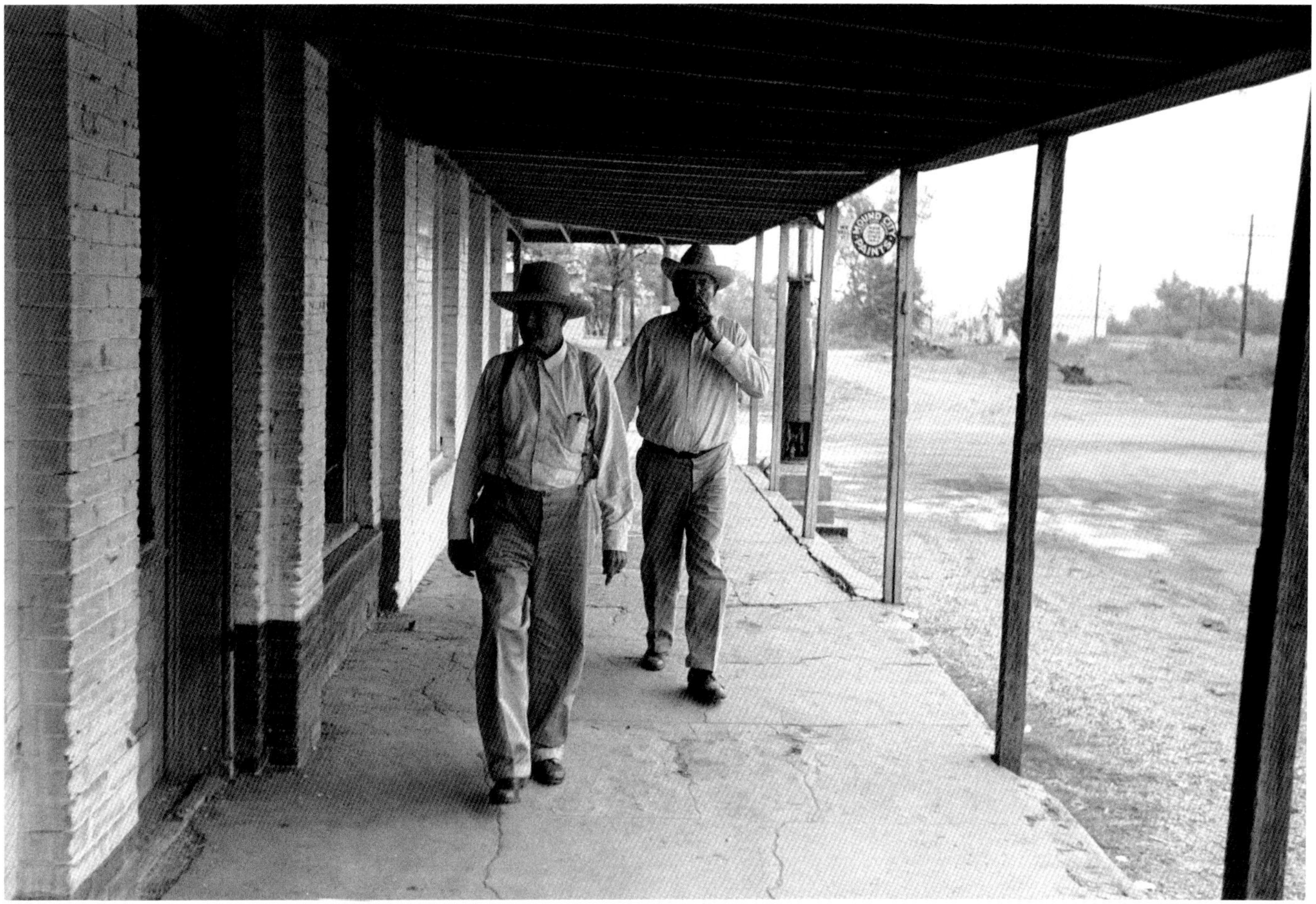
MOUND
PAINTS

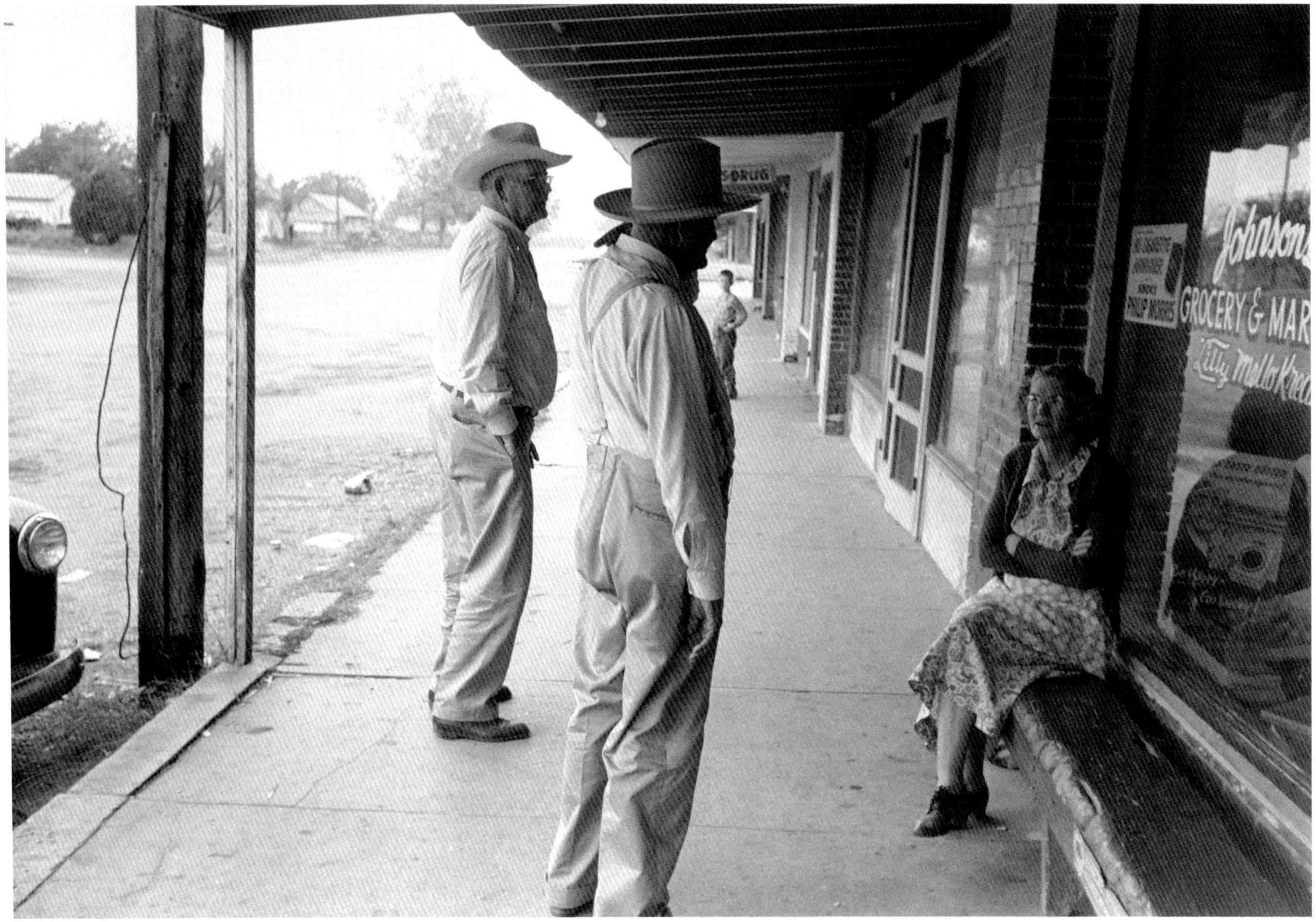
DRUG
PHILIP MORRIS
GROCERY & MAR

PCA Meeting

The Production Credit Association meeting isn't the place for work boots or a sweat-stained hat. Especially if you're on the board, like Hoyt. Then you wear a tie and a suit-coat. Not to impress anyone but simply to show respect. Respect for your neighbors and friends whose loan applications you must weigh-in on, whose loans you'll oversee, and sometimes, whose notes you'll have to call. To show respect for the job you've been elected to do.

When Hoyt was born in 1891, 80 percent of Americans lived outside its villages and towns and cities, most of them raising crops and livestock for their own consumption. The Industrial Revolution, built on iron and powered by steam and water, had begun moving people to towns and cities the century before. And now the Second Industrial Revolution, built on steel and powered by electricity and internal combustion engines, was in the midst of bringing

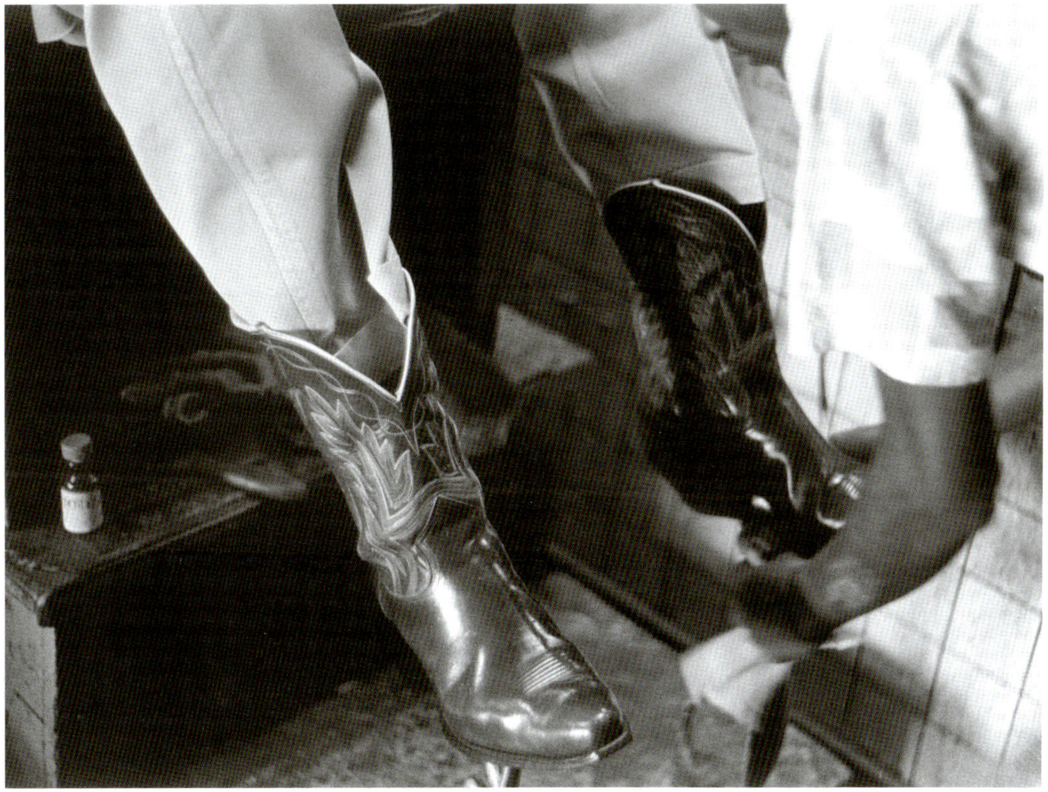

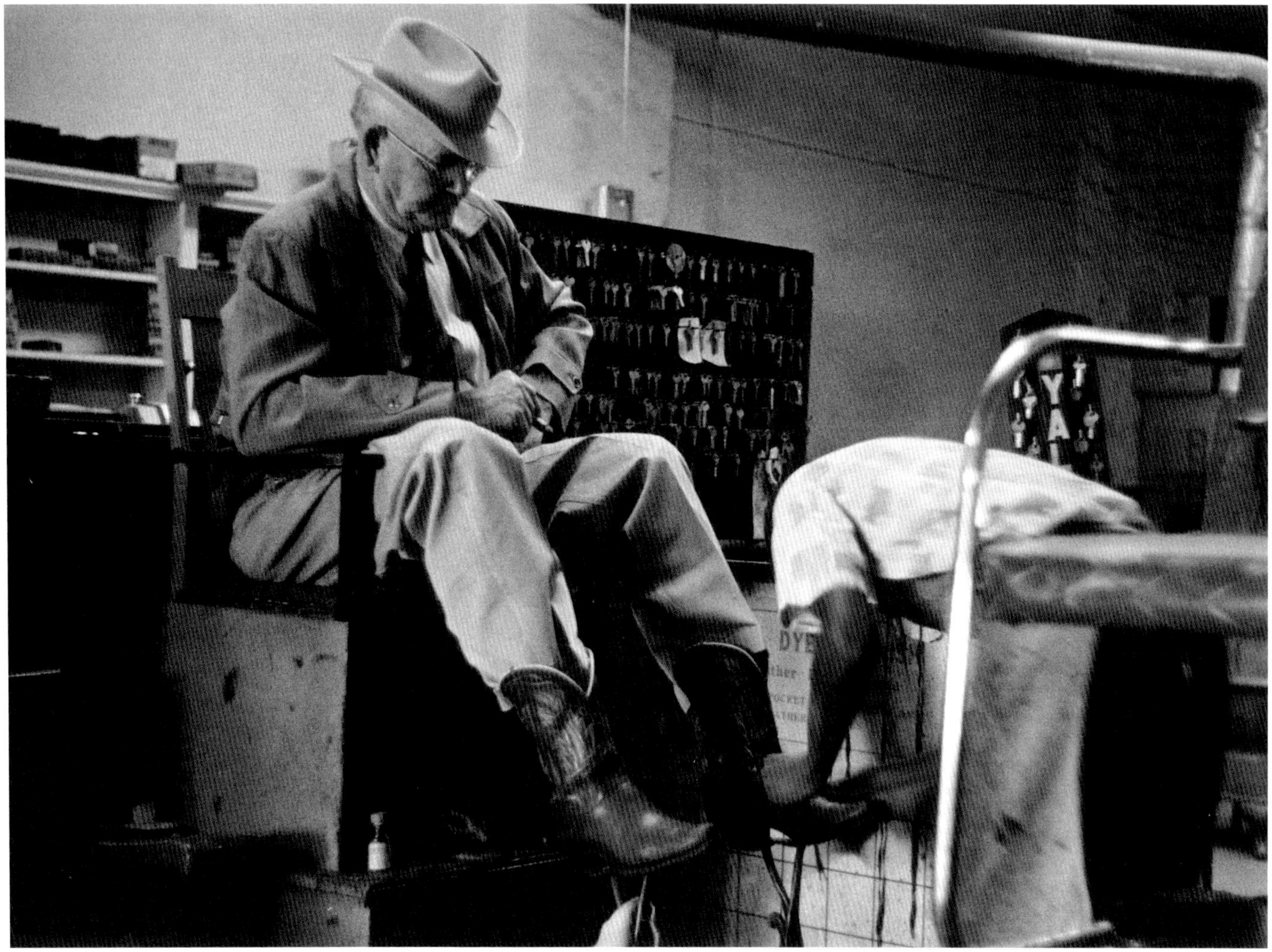

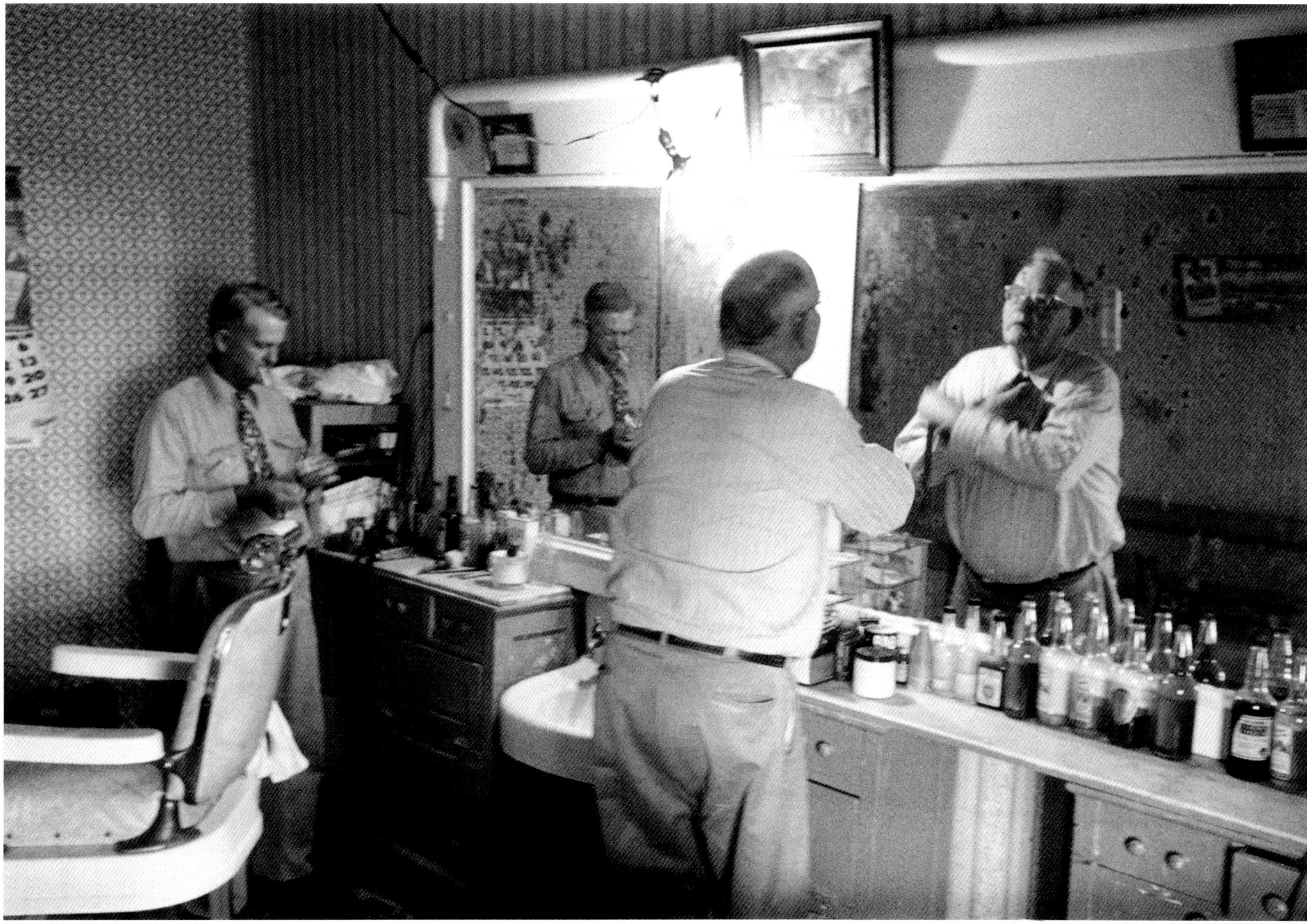

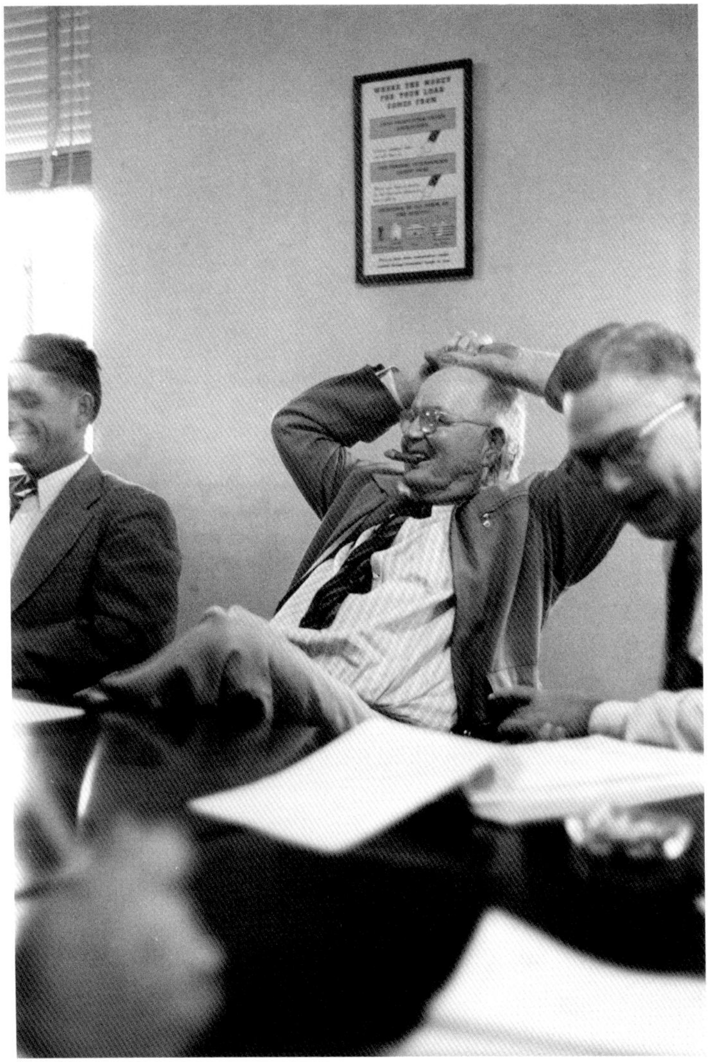

mechanization to the farm and laying the railroads to make it possible, and profitable, to ship the products of agriculture in to the cities.

A huge spike in demand driven by the First World War sped up changes on the farm, changes that required capital not readily available in the hinterlands. In 1916, the Progressives pushed through the Farm Credit System, organized on the co-operative model, to help small farmers and ranchers get the loans they needed to keep up with mechanization and the movement toward cash crops. When prices crashed after the war, a different kind of help was needed. In 1933, the New Deal brought the Production Credit Association to the Farm Credit System. National lending, local control.

It's a good day for a shine and a haircut. To show respect for where the money comes from.

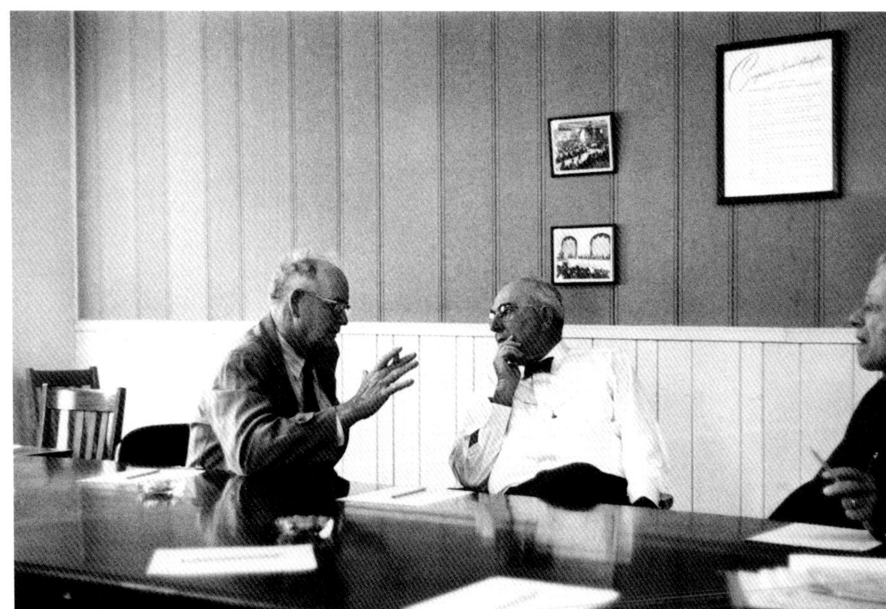

Sale Barn

Traders at the Chicago Merc buy and sell opinions. They go long on feeders for August delivery, maybe, or short on October live cattle. They make calculations in information. Supply and demand, inventories and stocks, the price of grain, all going up or going down, and when? Next month? Next fall? They calculate, but trust the feelings in their guts, hunches guided by a proverbial tribal wisdom: the best cure for low prices is low prices, the best cure for high prices is high prices, you can't go broke taking a profit, no one knows what it's worth 'til the day it ships.

And today it's shipping. Opinions meet facts in the sale barn. Not everyone here is a buyer, or a seller. But everyone's a trader. Like it or not, we're all in the market. And today, the market's not numbers and symbols on ticker tape, not reports in the *Livestock Weekly*. Walk through here and the market gets on your boots. It's in the auctioneer's chant, in the ground-man's sharp lookout for the next bid, shouting "eh-Yup!" when he sees it, throwing his arm high in the air, carrying the price per hundredweight up with it. While the money's hot, the market's in the faces under the hats, all trained on the ring, focused and intent.

Later there'll be time for iced tea and talk, for conversation and coffee, for post-market analysis. Everyone needs information. Everyone checks their gut.

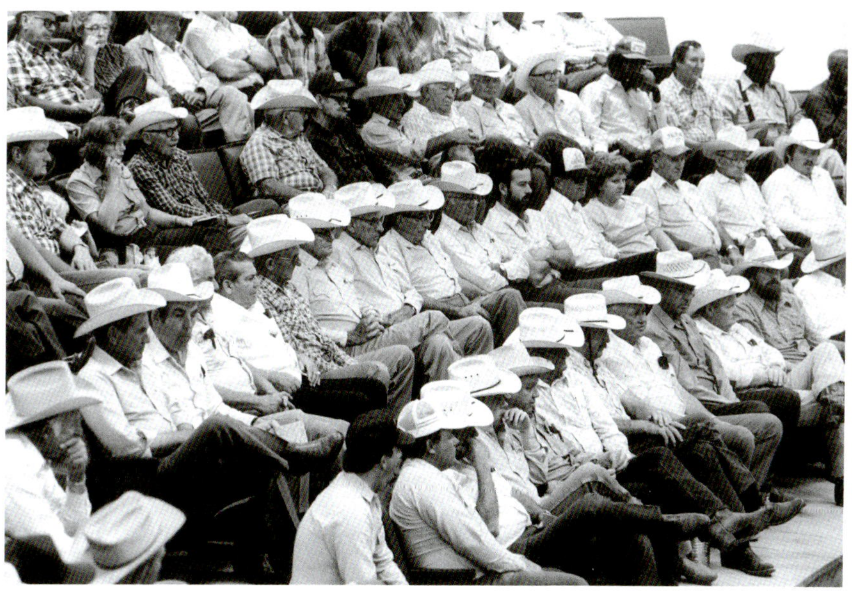

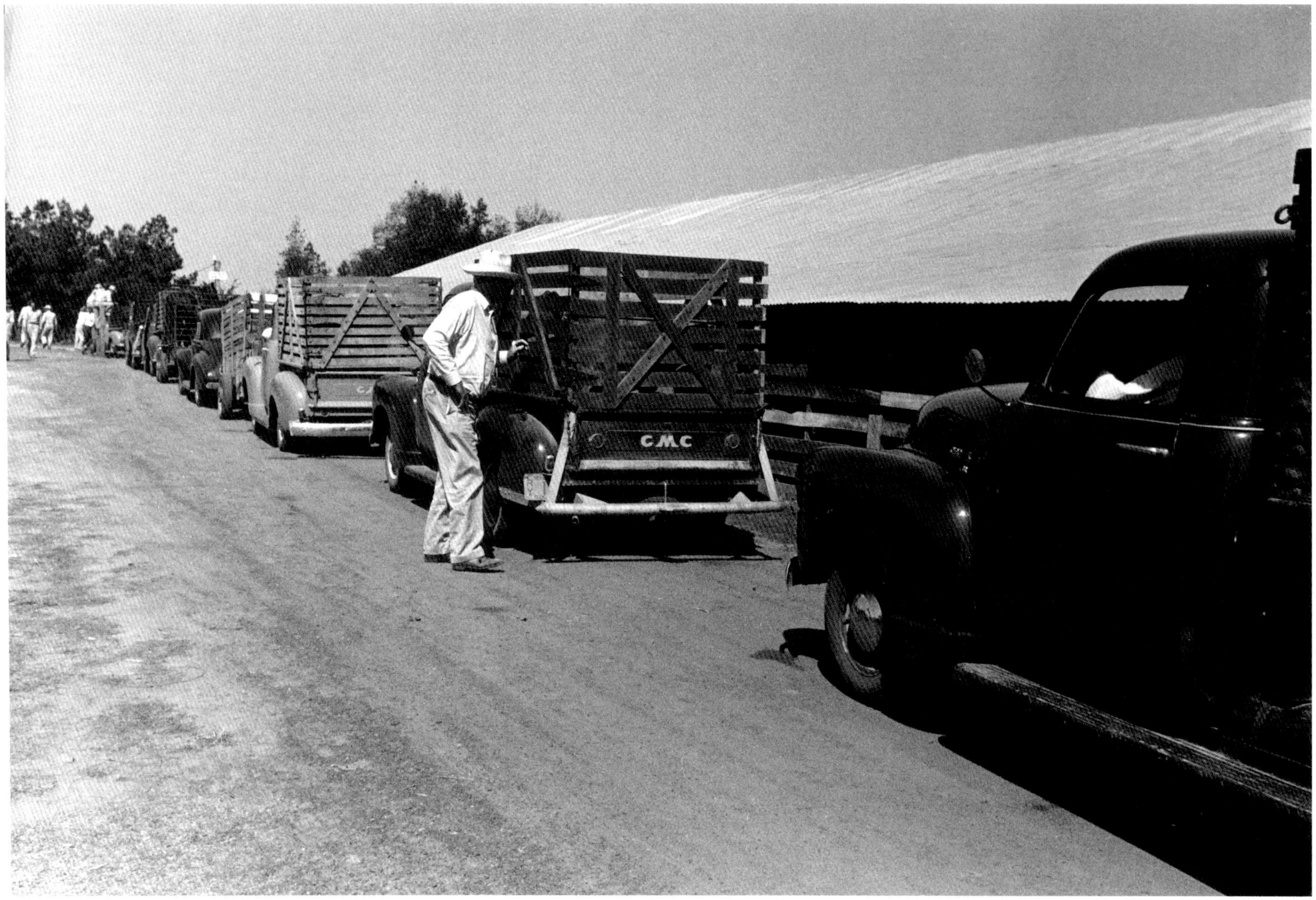
GMC

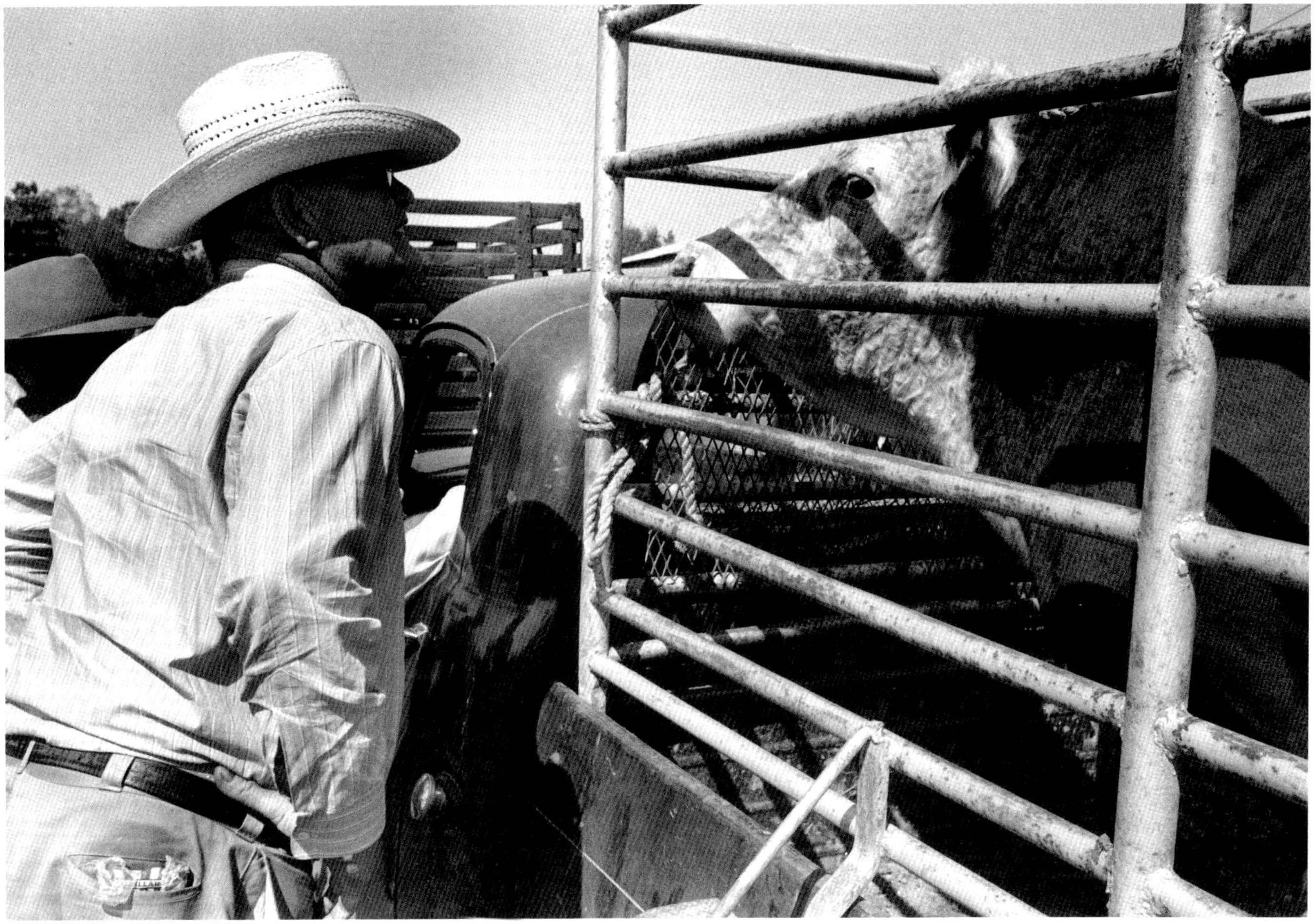

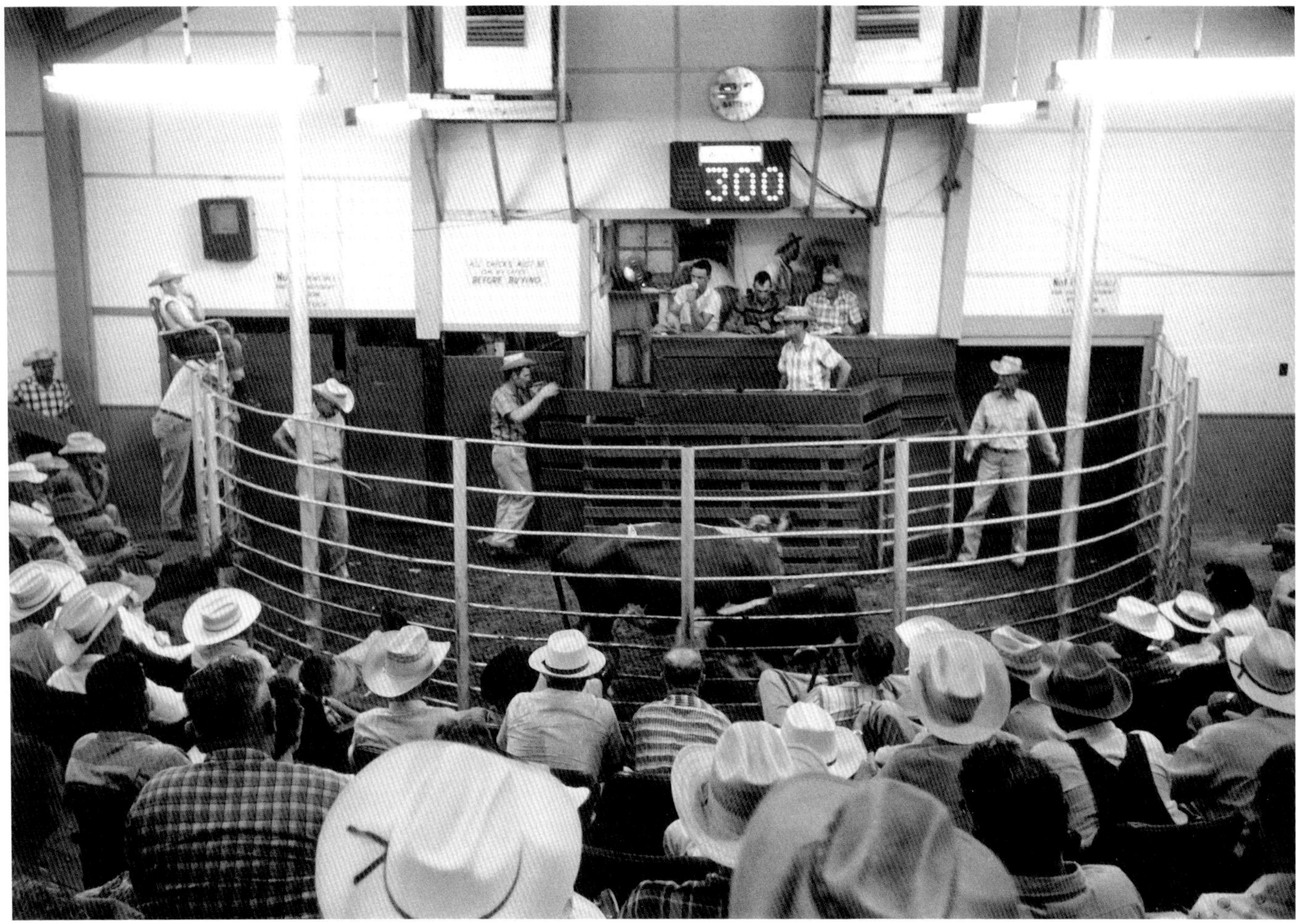
300
BEFORE BUYING

Branding, 1951

The world was first cut up by farmers. Raising crops demands rootedness, the kind of year-round attention to a specific piece of land that divides the world into mine and yours. Farmers are born of fences, and fences are born of them.

But confinement is antithetical to herders. Cattle require grass, and lots of it. Grass, in turn, requires cattle, flourishing in a cycle of intensive grazing followed by long rest, a cycle that insists upon ample ranges, making nomads of both cattle and herders.

Still, there is yours-and-mine. On the open range, we must be able to separate whose is whose. Even among family and friends. Among others, we must always be able to mark what is ours, no matter that the open range has been fenced. At least since the ancient Egyptians, the mark has been made by branding.

Romans branded, too, as did medieval Spaniards. The Romans, we must think, brought the practice to the Celts, the Celts to the tribes that became English and Scottish and Irish and Welsh, and the descendants of those tribes and the descendants of the Spanish brought branding to the New World with their cattle.

The cycle comes full at the Porter Place. The range is assembled from farms, the nomads rooted by rotational grazing. The herd is branded with the Saddle Stirrup, the hot iron wielded by the progeny of the ancient cattle cultures, Hoyt and Claude Searcy from the English and the Scots and the Irish, Dayton from the Nubian princes of Egypt.

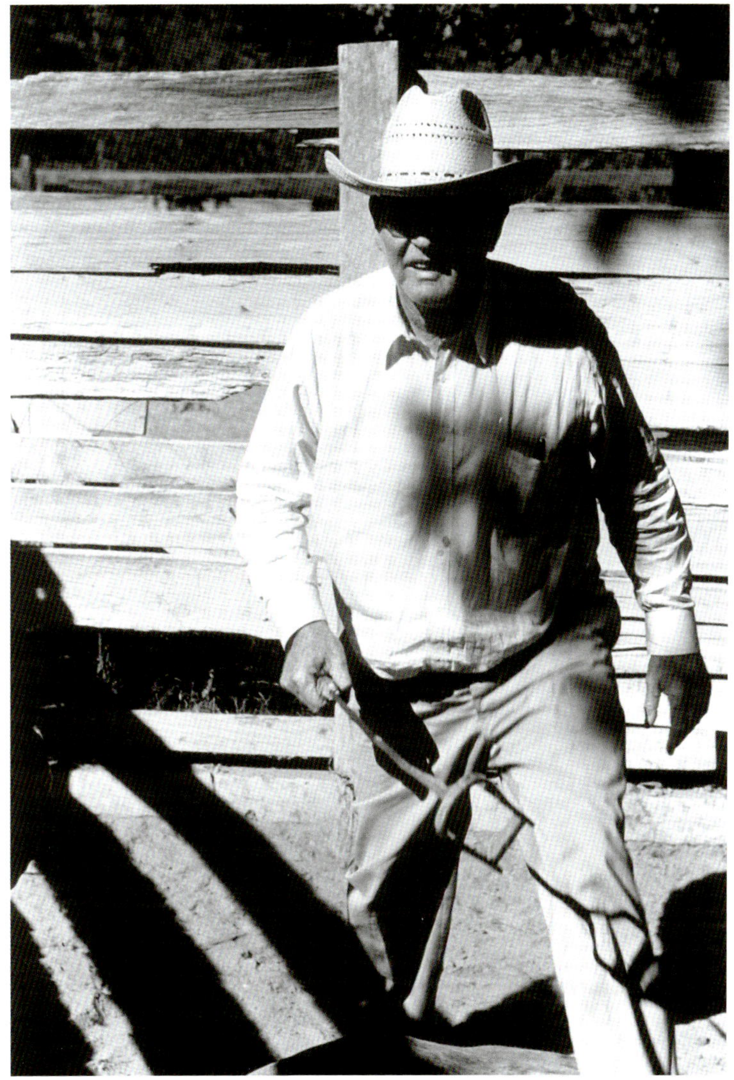

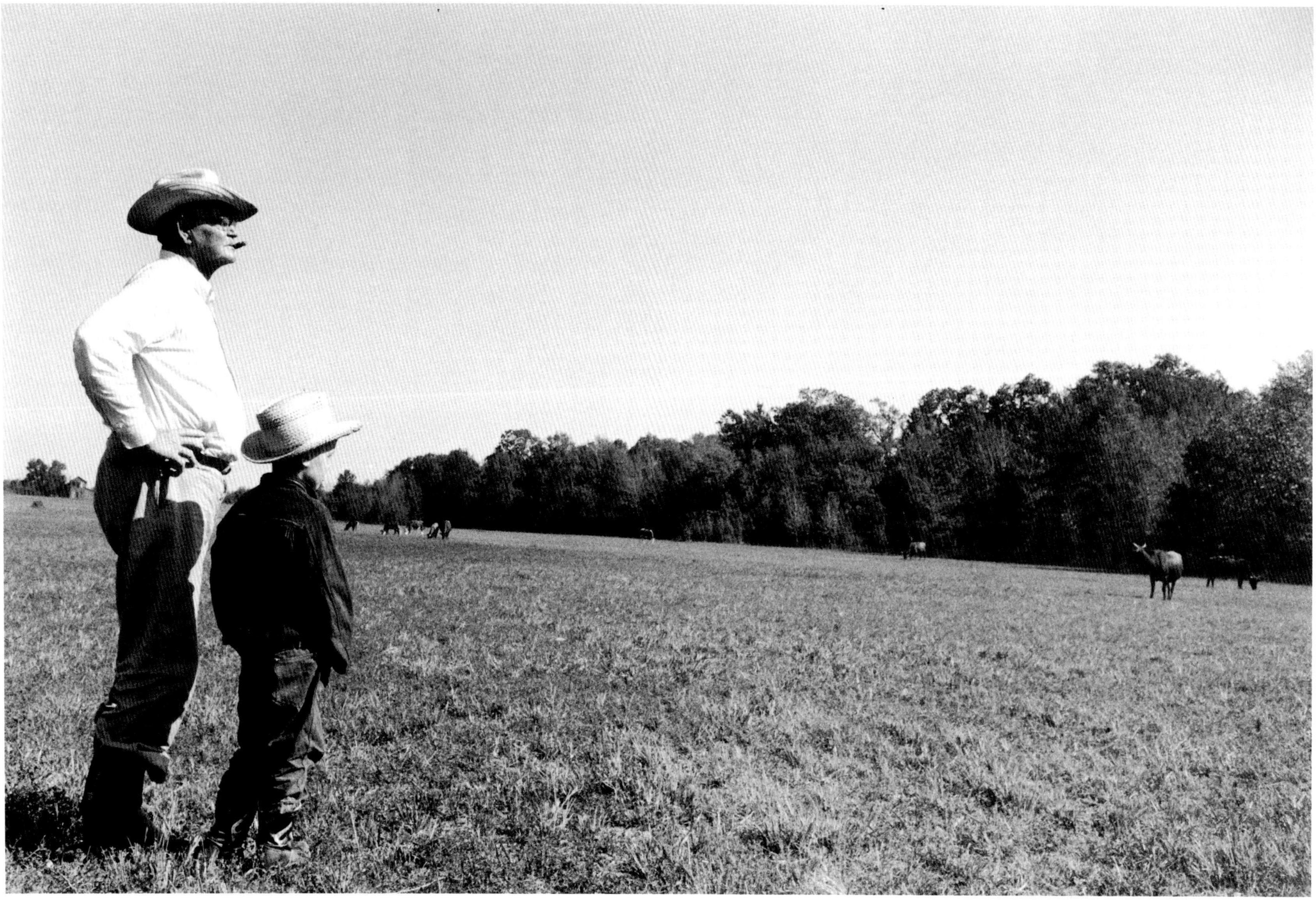

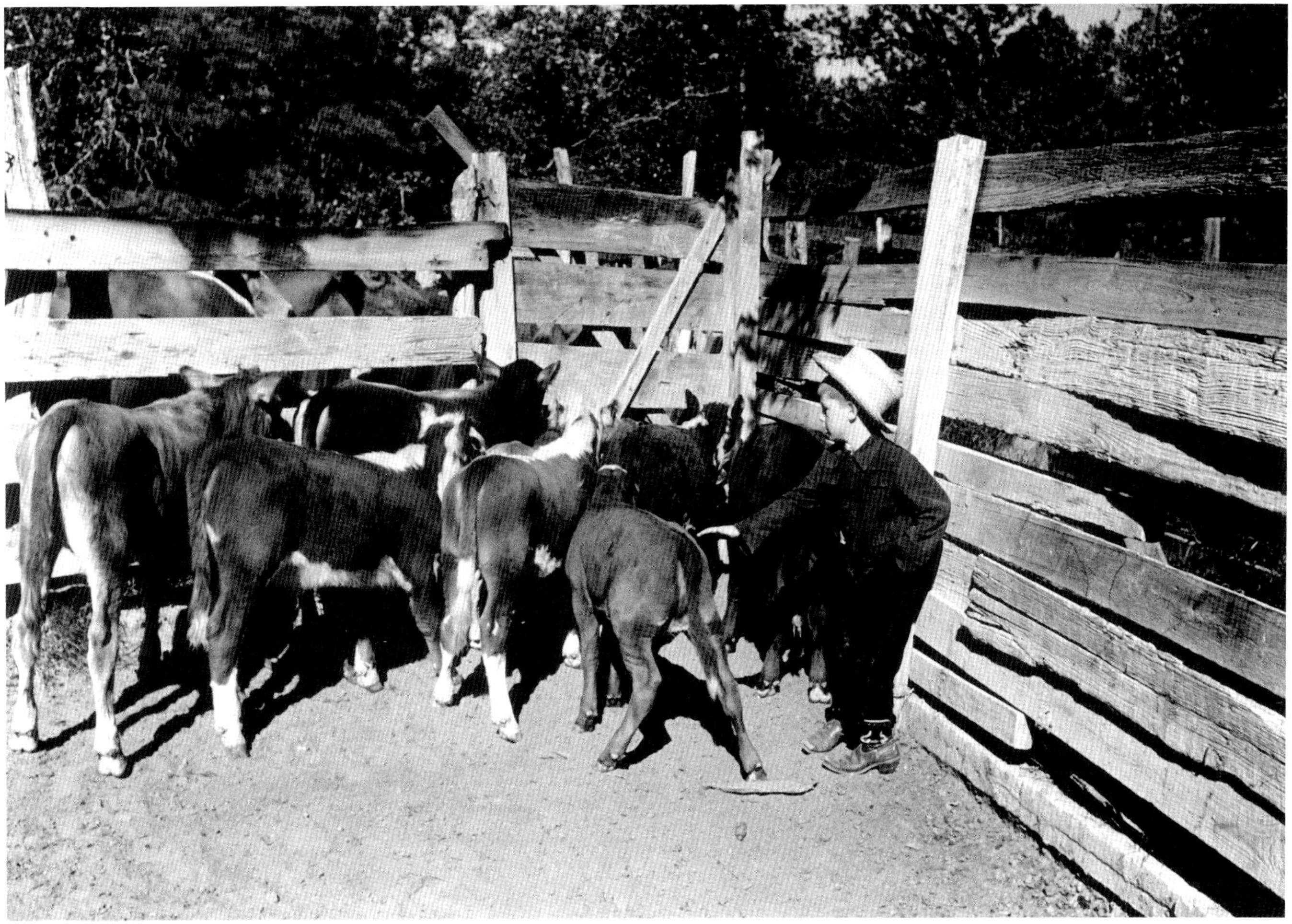

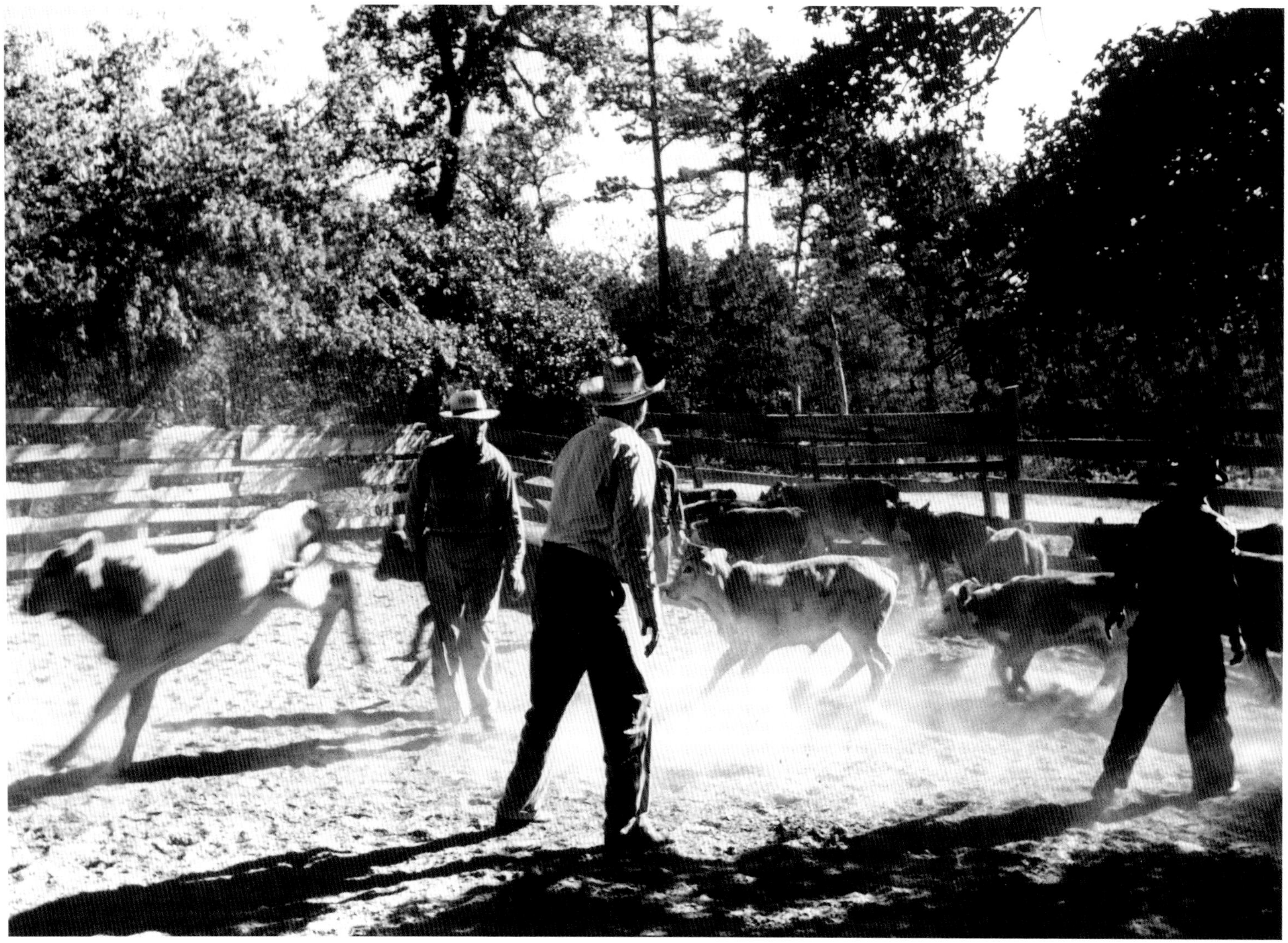

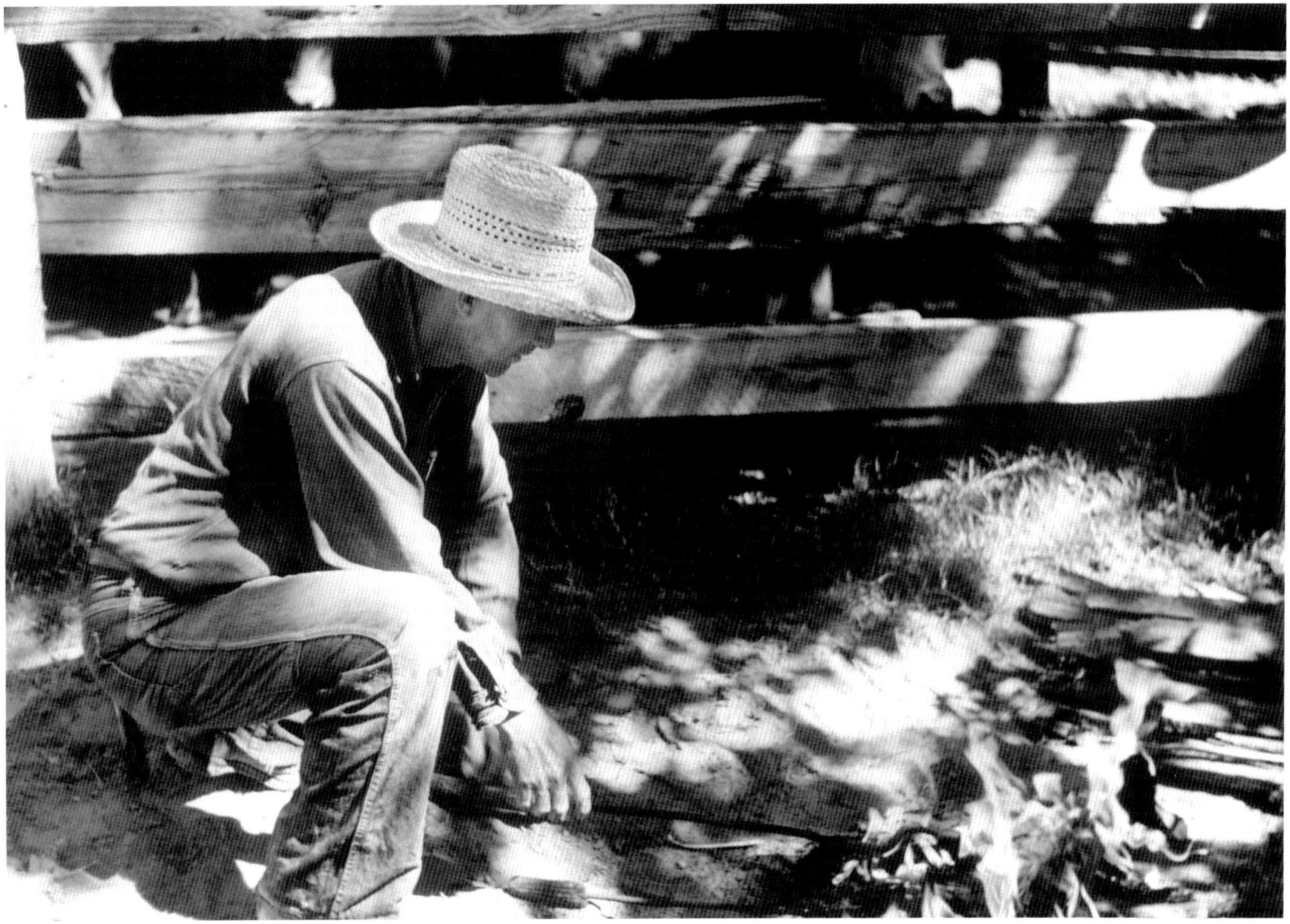

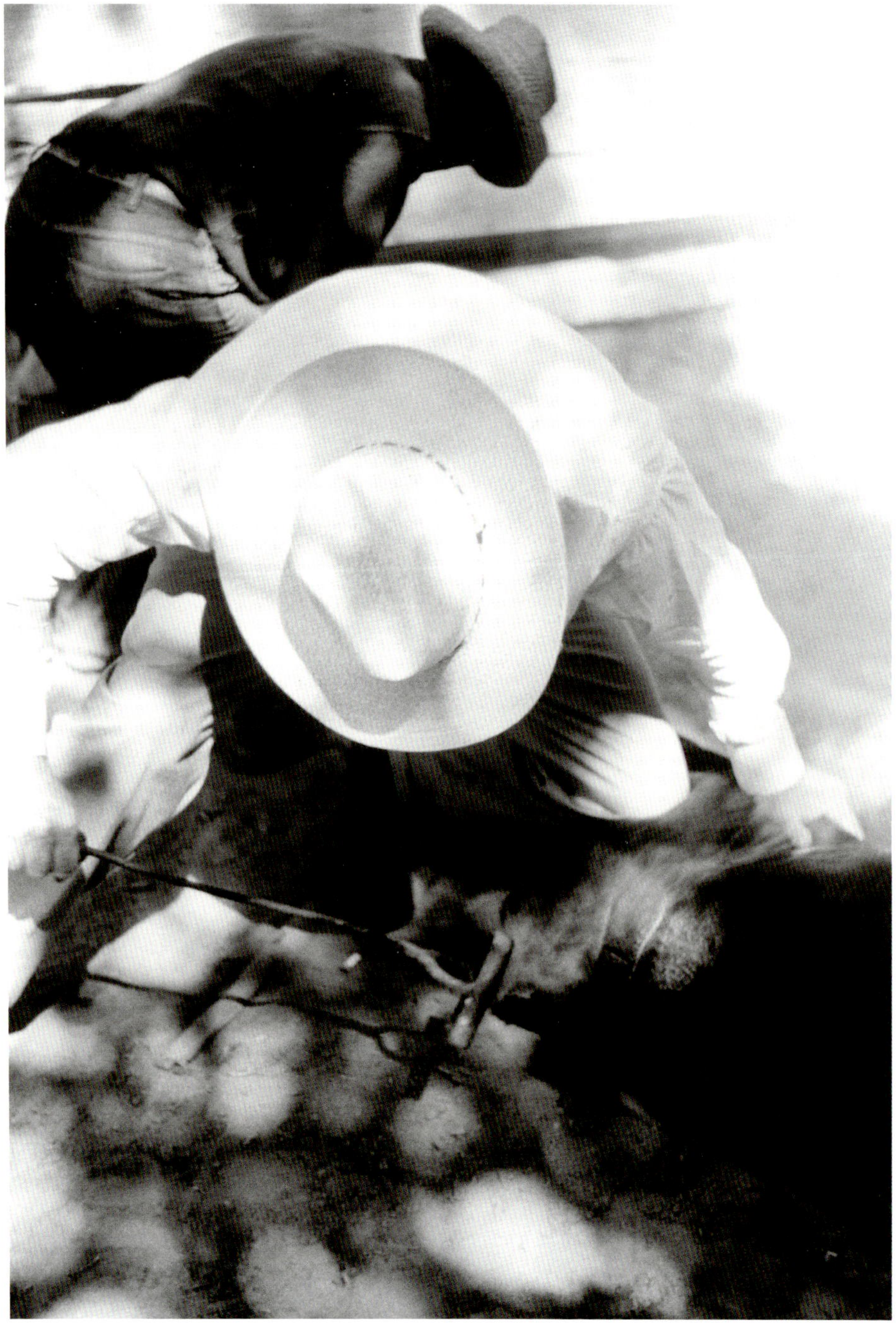

. . . even with your brother, smile but bring in a witness. Trust and mistrust alike have ruined men.

HESIOD, *Works and Days*

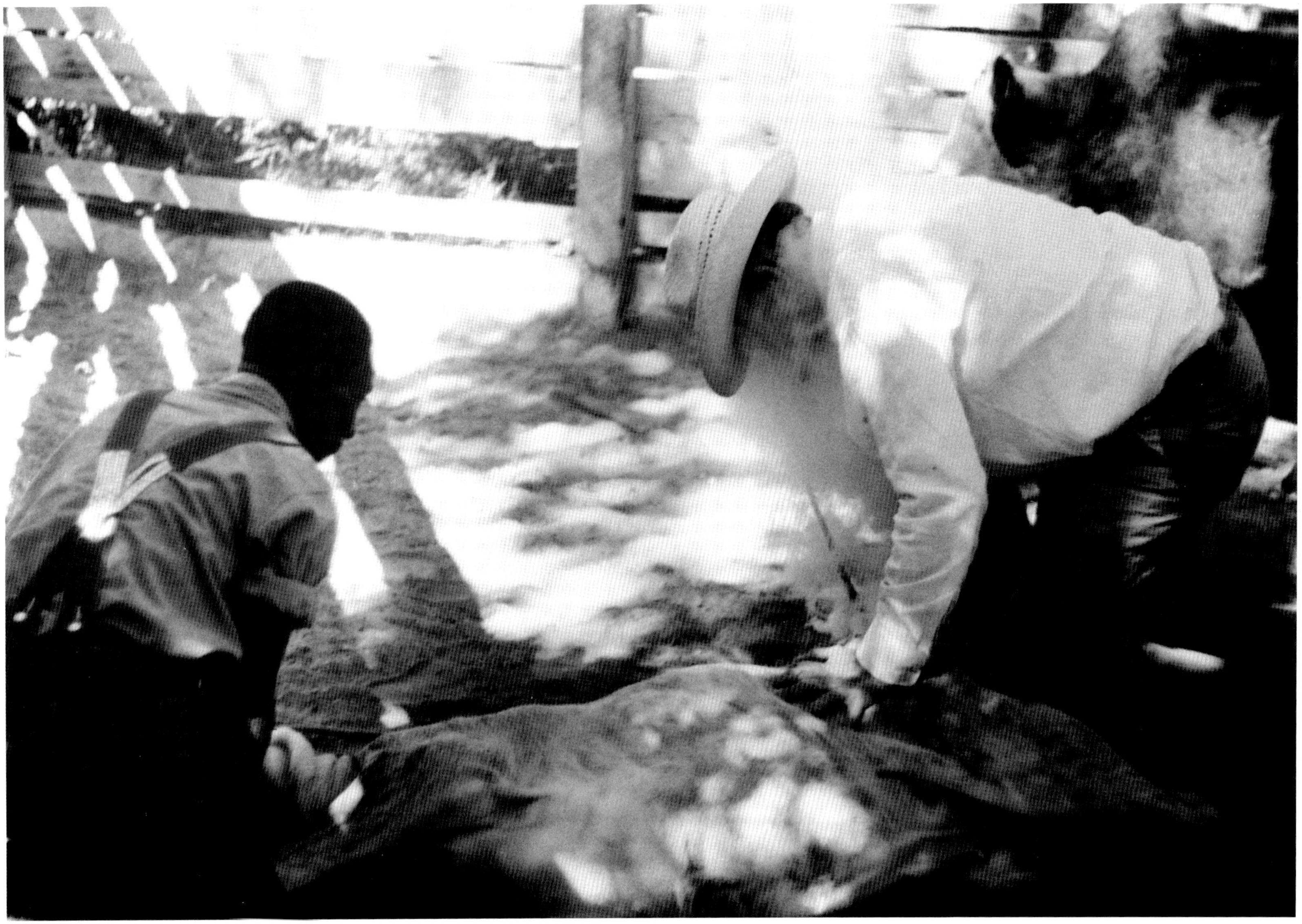

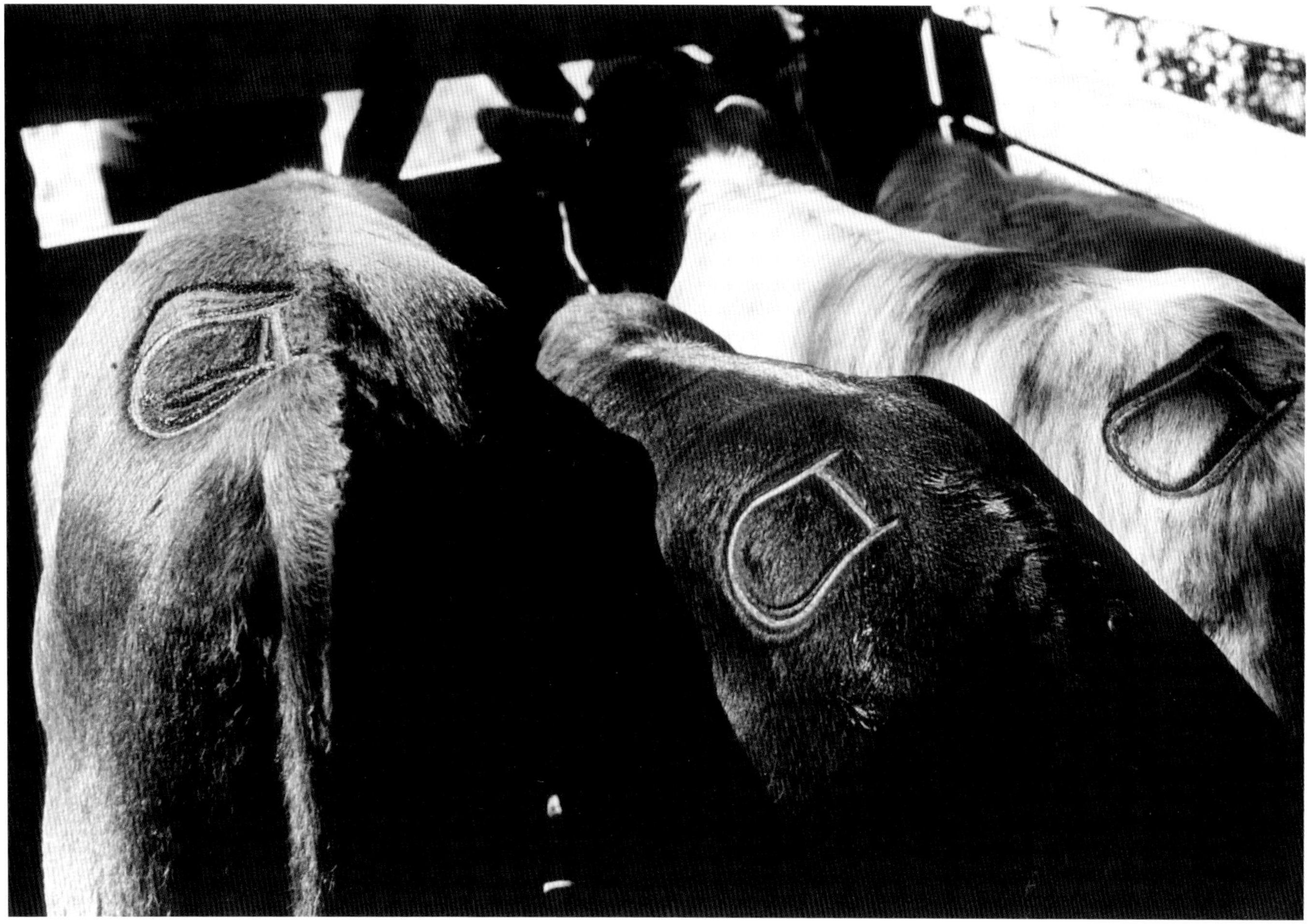

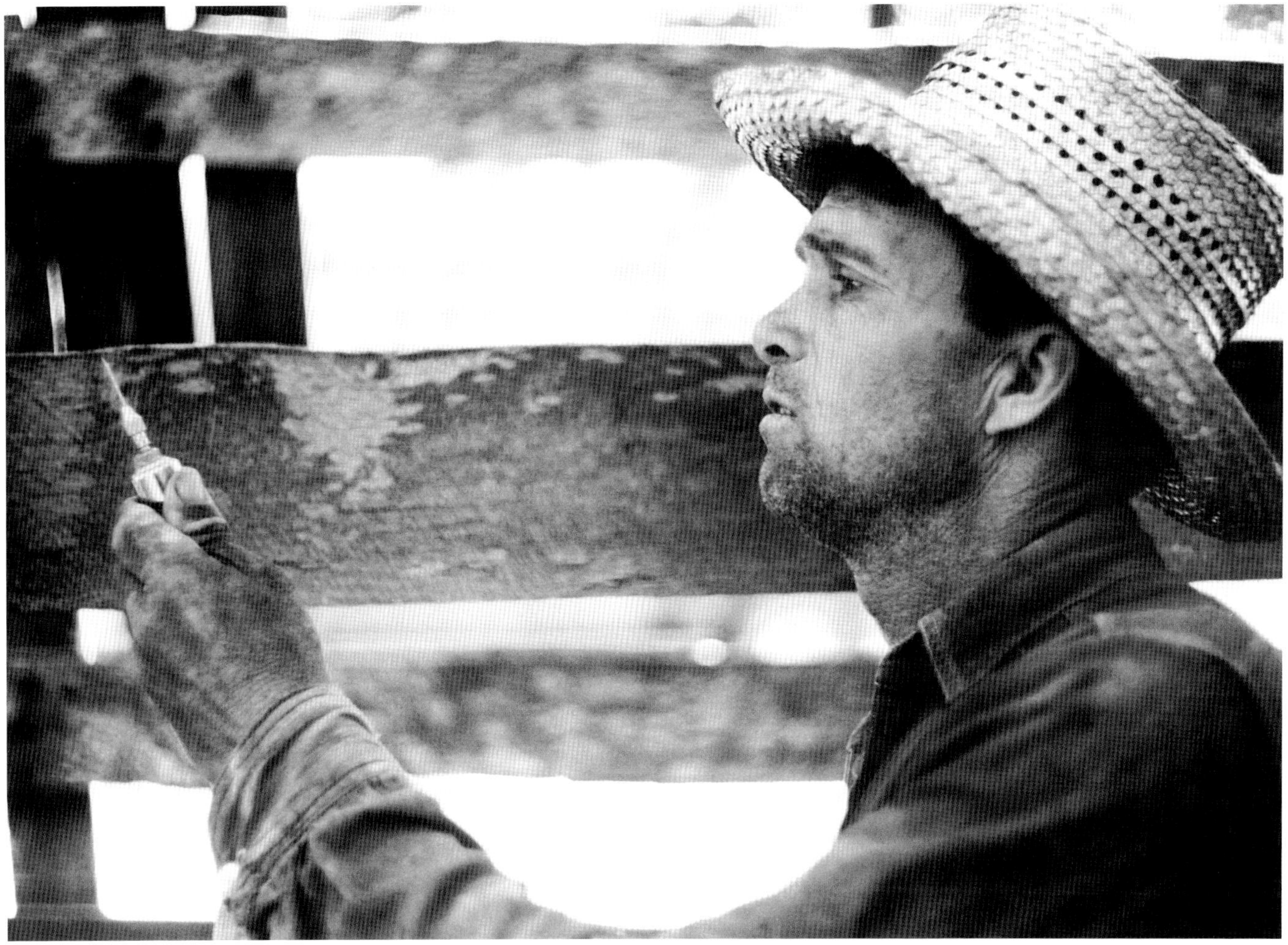

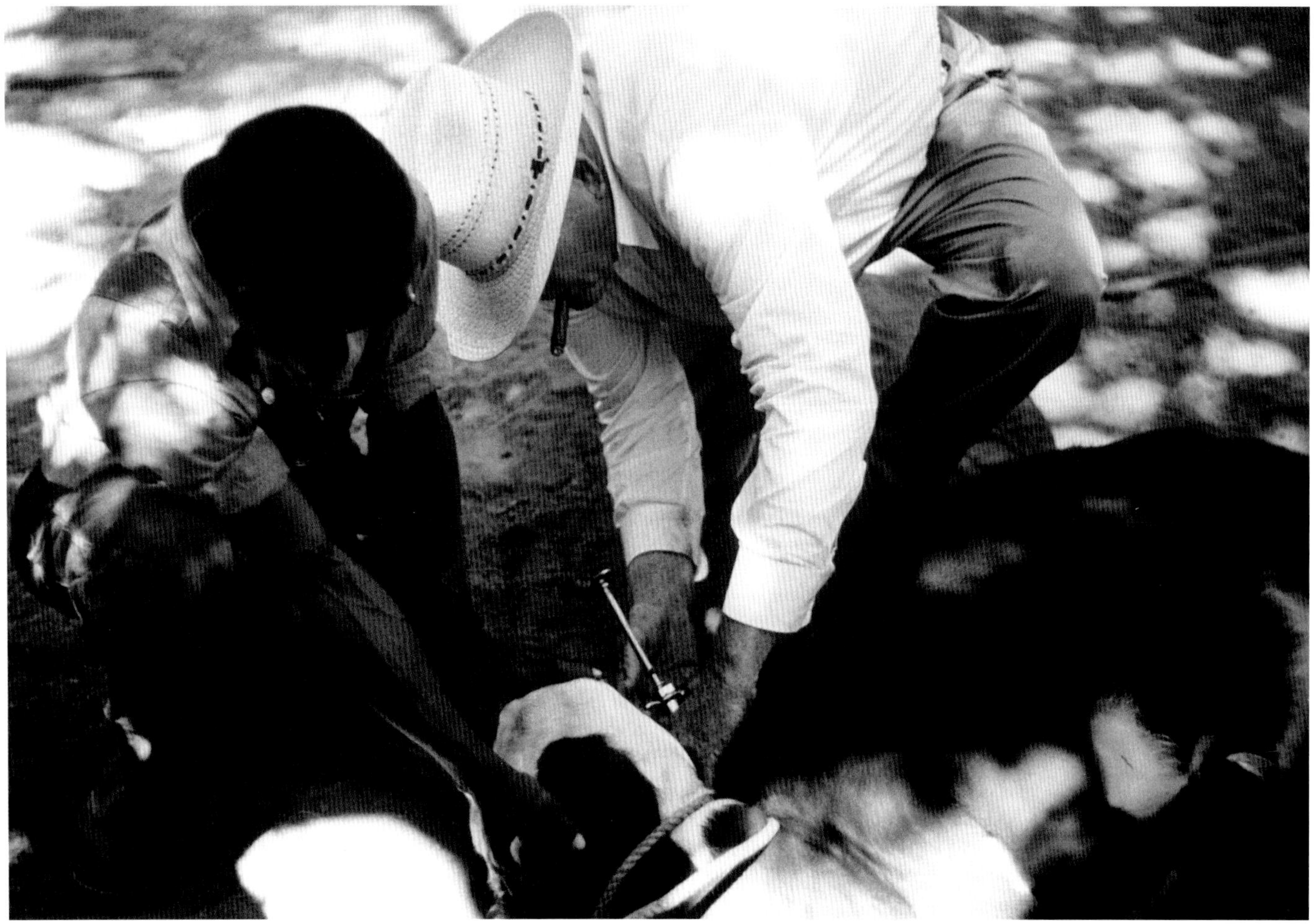

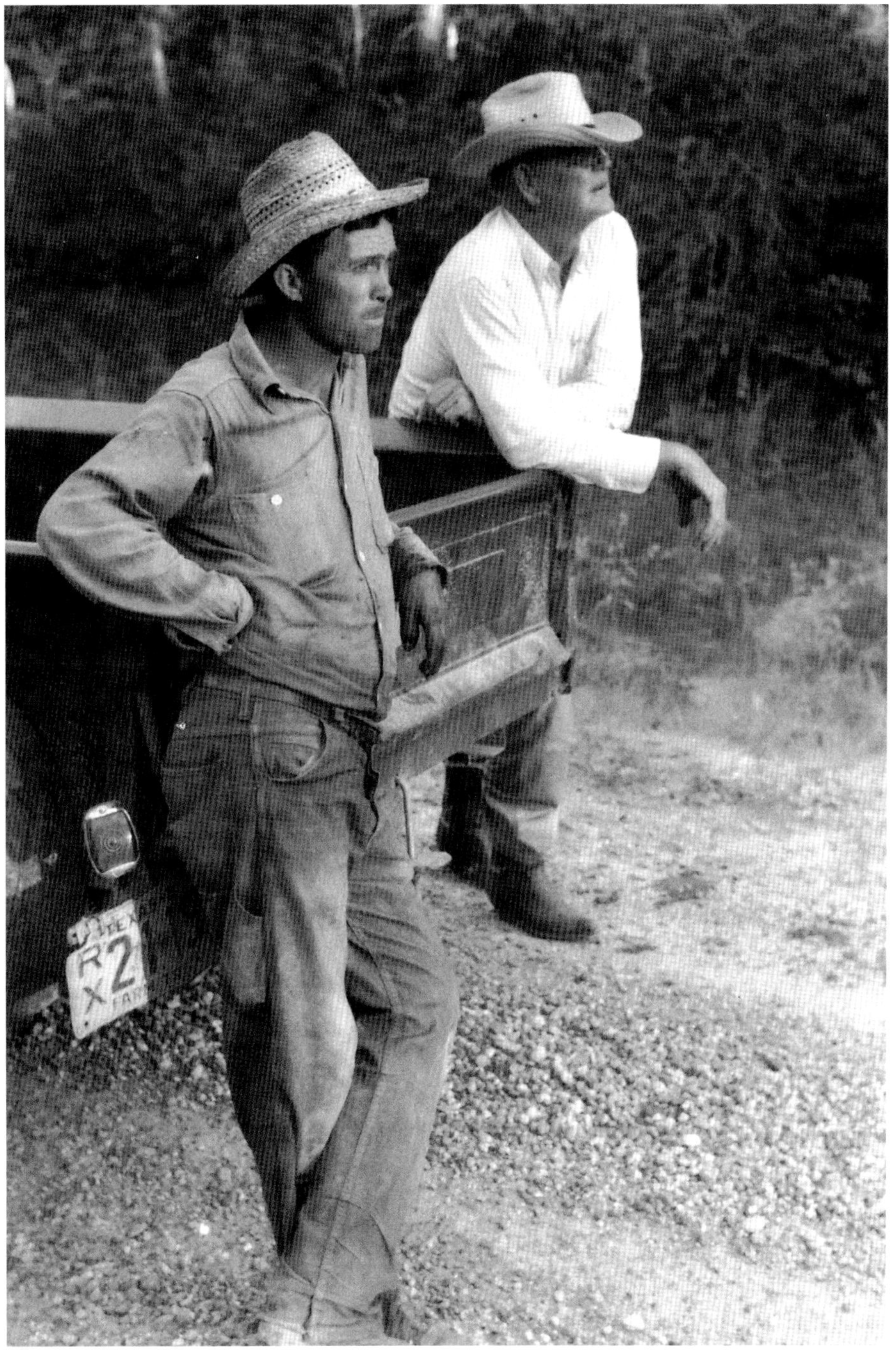

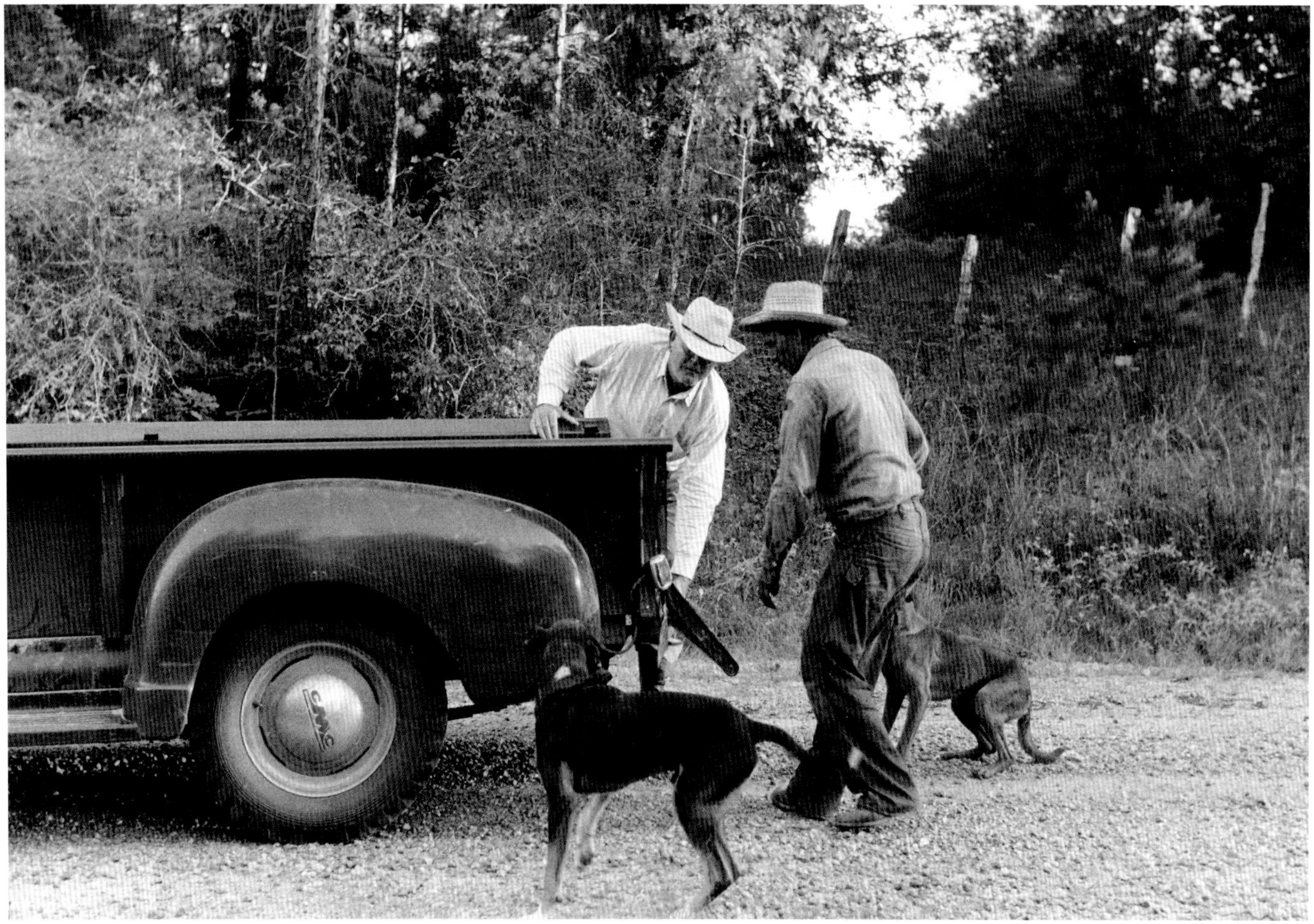

Wounded Cow-Dog

The Porter Place is where the South meets the West, where post oak savannah and piney wood forests drift into open prairie, where pastures are carved from dense hardwood thickets, where cattlemen ride western horses and use southern cow-dogs. Further west, dogs are heelers, bred and trained to work cattle by pushing them from behind. Southern dogs work differently: from the head, baying-up the cattle, holding them in bunches. And they're bred differently, mostly hounds, like the Blackmouth Curs favored on the Porter Place. As with so many things southern, they're multipurpose, too, used for hunting hogs and treeing everything from 'coon to bear.

Working cattle is dangerous for any dog, hound or heeler. A cow will fight, charging and kicking, using horns if she has 'em, always using the sharp hooves. A cow-dog has to be tough and aggressive on the cow, smart and obedient for the cowboy, protective and faithful to its place and its people. A dog like that, a good cow-dog, is worth a lot. Sometimes as much as several hands. Sometimes more.

So when a cow broke Red's leg, Hoyt didn't think twice about carrying him to the vet. He took his grandson, Guy Porter. Blackmouth Curs are good with kids, and the bonds are strong, even when the boy comes only for the summers. A hound like Red knows who belongs to the family, knows who belongs to their land.

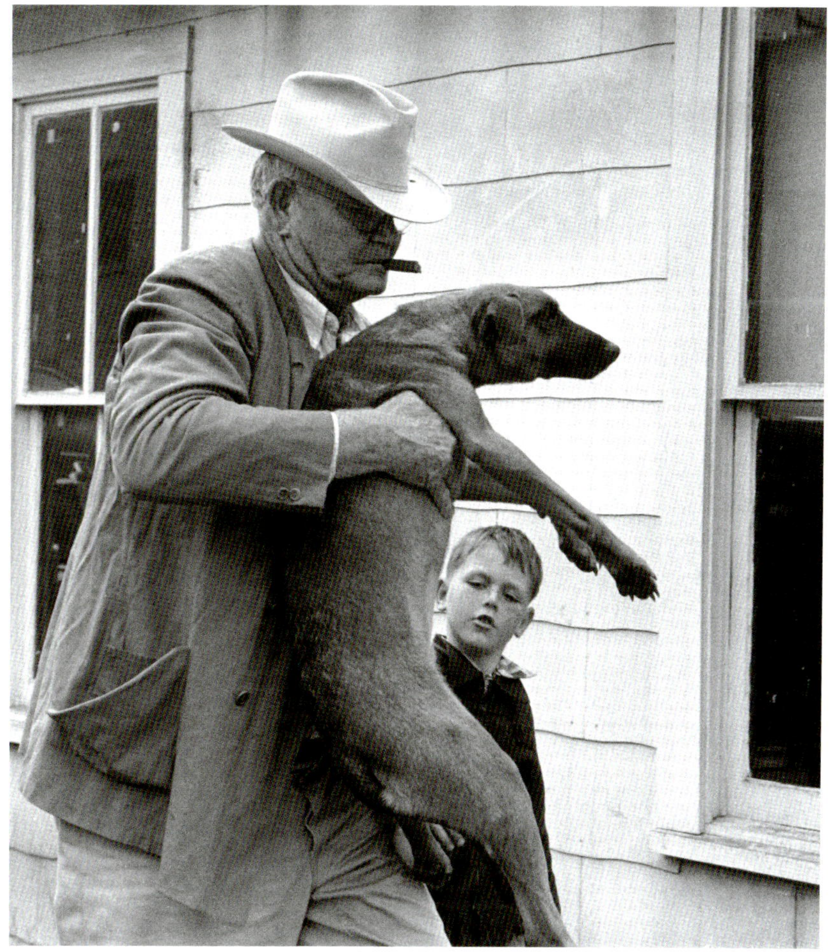

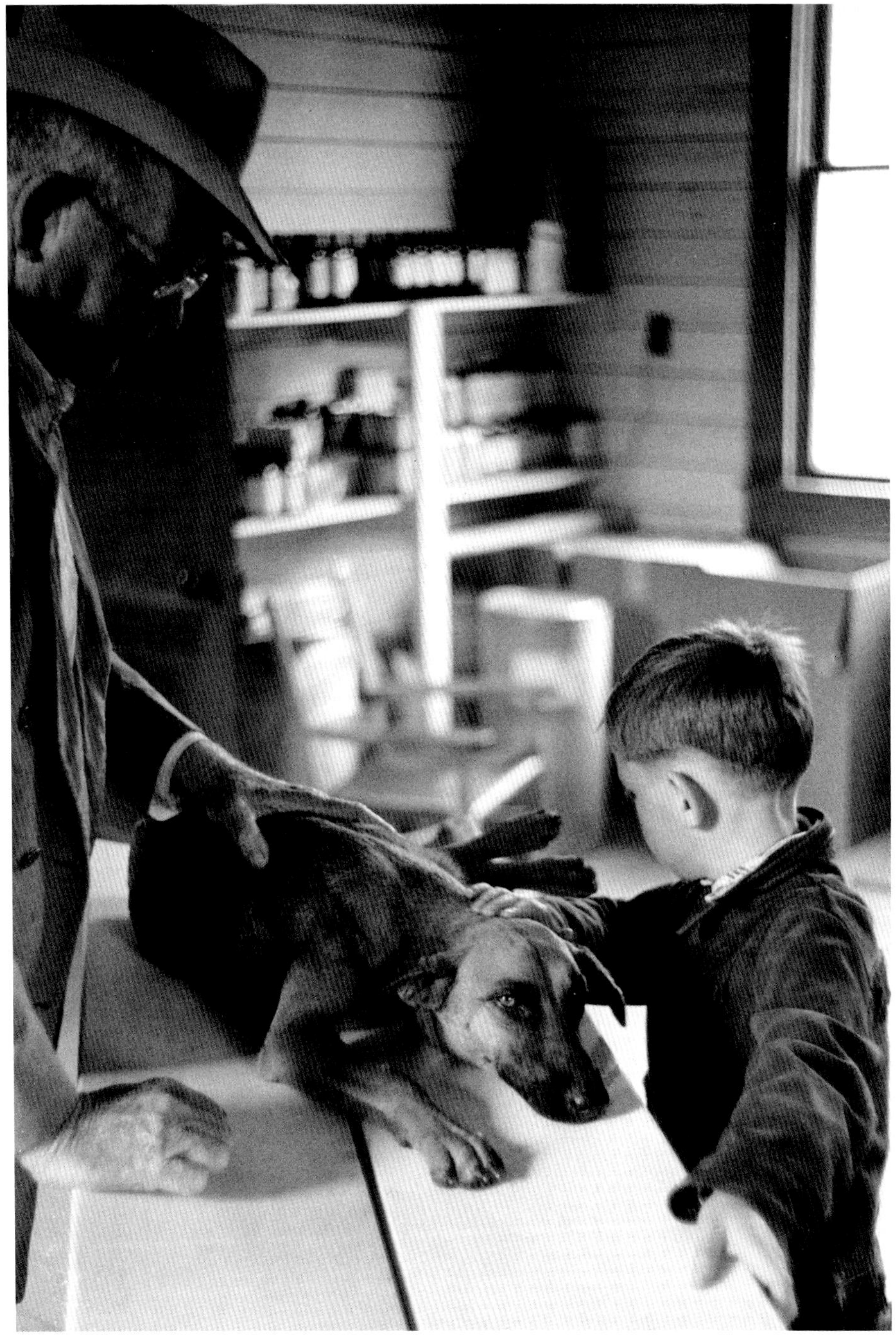

Guy Porter, no more than six or seven, stays by his dog, unflinching, even when the doctor pushes in a big needle to deaden the leg before setting it. In the end, the vet's work is good, good for both dog and boy. Hoyt can bring Red back to recuperate, to work another day.

It was rumored that one photograph of this series — the young Guy Porter leaning against the table, his head sideways on his shoulder, his hand holding his dog's head and his eyes deep into the eyes of his pet — it was rumored that this is the only photograph ever to move Edward Steichen to tears. We can't know for sure. But then, these are not snapshots. These are careful documentary photographs in which Gillette's composition directs our eyes to the only eyes fully shown in the images, to the eyes of the boy and the eyes of the dog. Remember "Old Yeller"? He was a Blackmouth Cur, too. Maybe Steichen did indeed cry.

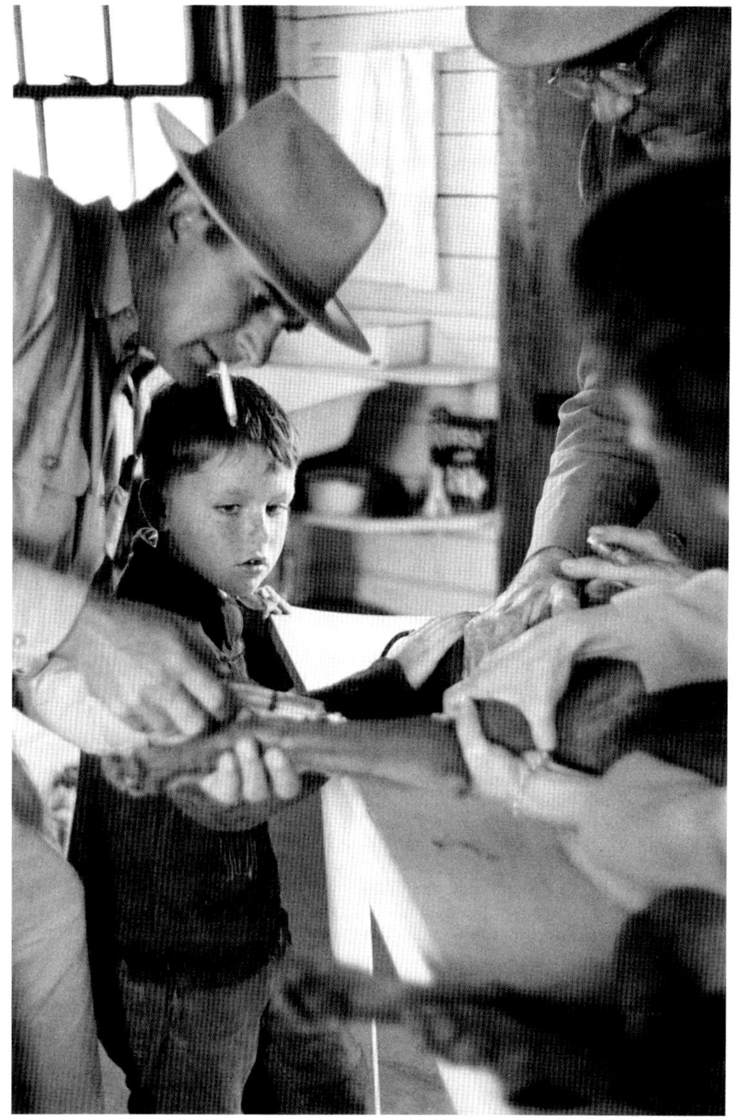

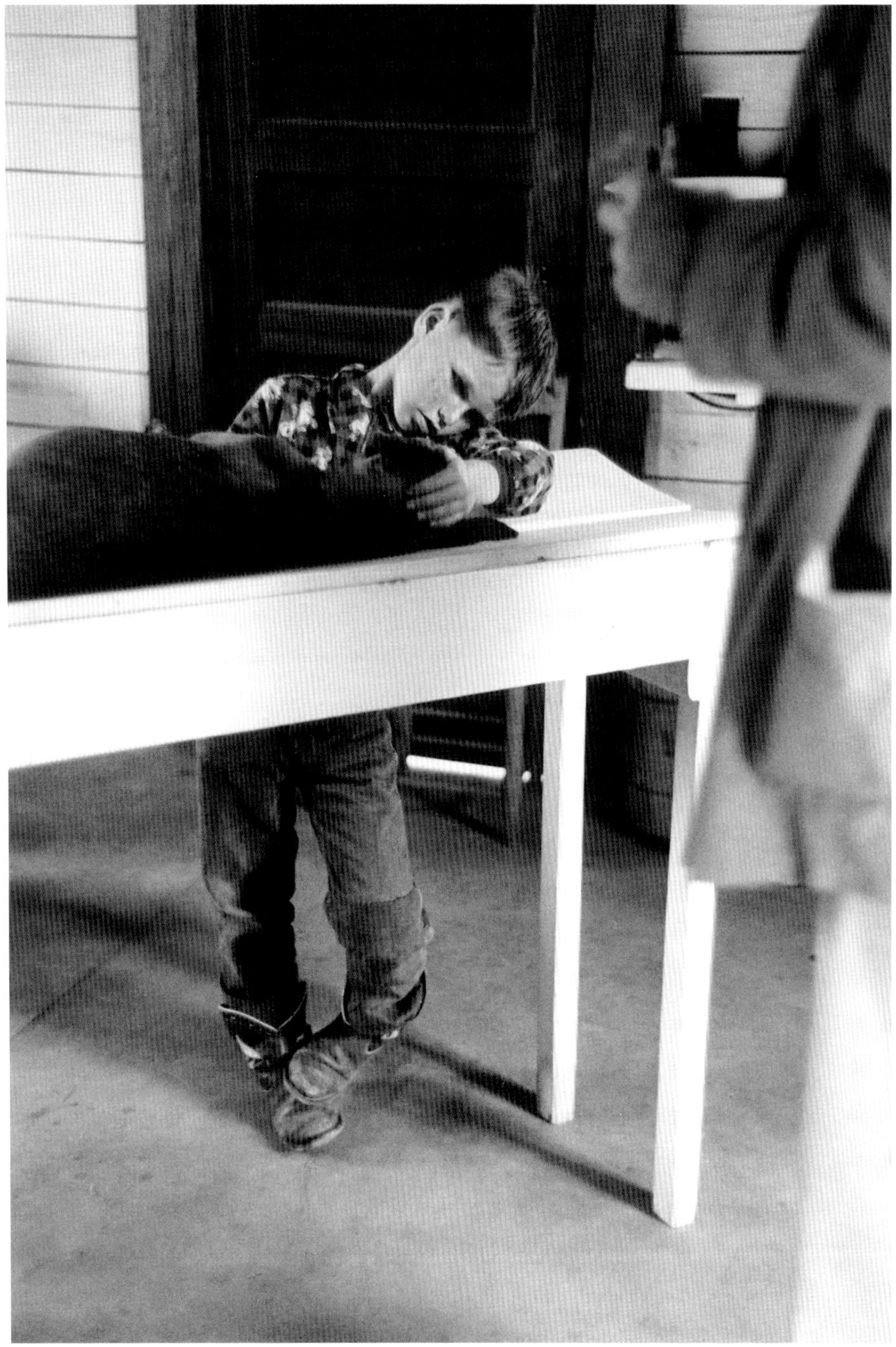

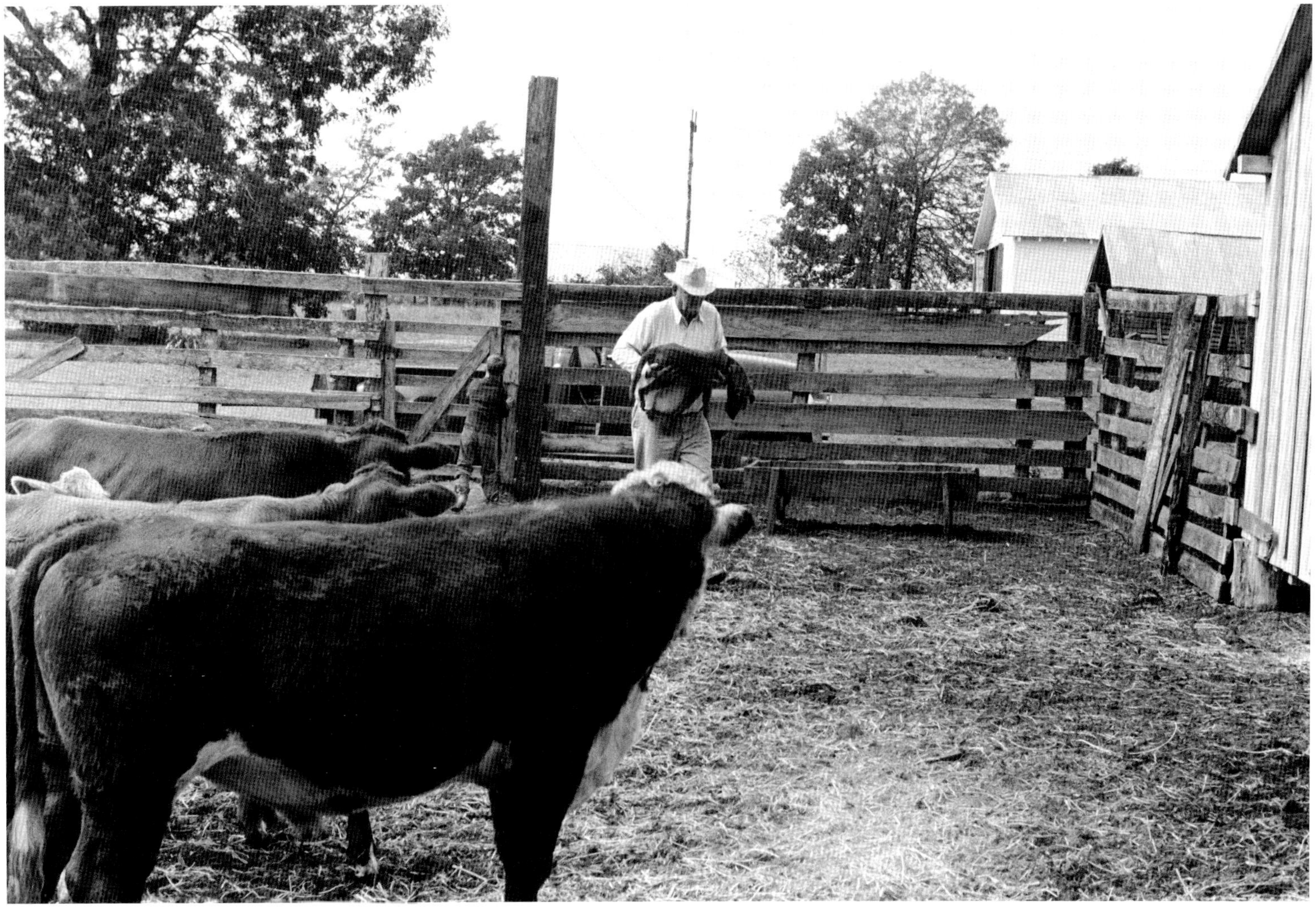

With Dayton

The naming of things. It is perhaps the only activity that is only human. Whales sing. Dolphins speak. Bees dance. Ants build, not just a hill or a mound but whole societies. Birds make tools — and make tools to make tools. All manner of animals play, hold grudges, learn from experience, and pass along what they learn to their young. They mate for life or mate indiscriminately. They are courageous and cowardly, noble and mean, faithful and treacherous, loving and hateful.

We are all that. But we name things, too, and we name them so easily and so naturally that the power of a name most often escapes us. For naming has consequences. Naming the unknowable, the mysterious, this way of naming gives us religion. Naming what can be known, mysterious or not, gives us philosophy. Naming what is known gives us science. Knowing what can never be named gives us art.

These ways of naming give us power, they grow us beyond ourselves. Other ways of naming limit us, make us smaller, create false choices. Black-and-white, or color? A comparison in a question, one name that narrows, one name that expands. One name that is only two things, one name that is infinite things. To understand the consequences, study these photographs. We name them black-and-white, but nothing in them is ever fully black, ever fully white, but always the endless gradations of tones between the all and the nothing. There are those who will let the narrow and limiting name stand for the expansive and infinite thing.

But not Guy Gillette, who chooses to work in black-and-white, not in color. Not for expediency or simplicity. Not to reduce his choices, his means, his outcomes. Rather to expand them by concentrating his energy, and ours, on what is important. And what is important in documentary photography is fact, and fact is conveyed in an image by composition, and composition is built of tone and line. For while we all see tone and line in the same way, we all see color differently. Color is perception, color is feeling, not fact.

White, or colored? Another comparison in a question, another name that expands, another name that narrows. Another set of limits creating false choices, choices that make us smaller. To understand the consequences, study these photographs. In tones of gray, these images are never white or colored, never asking that question, never making that comparison. The lightest gray tones are in the sunlight falling on shirt or hat brim or blaze marking, the darkest of gray tones are in the shadows under the brush, and the middle gray tones are in Dayton's face in the pickup bed and Hoyt's face in the pickup cab.

The film is black-and-white. The photographs are not. The people in them are not. The naming of things.

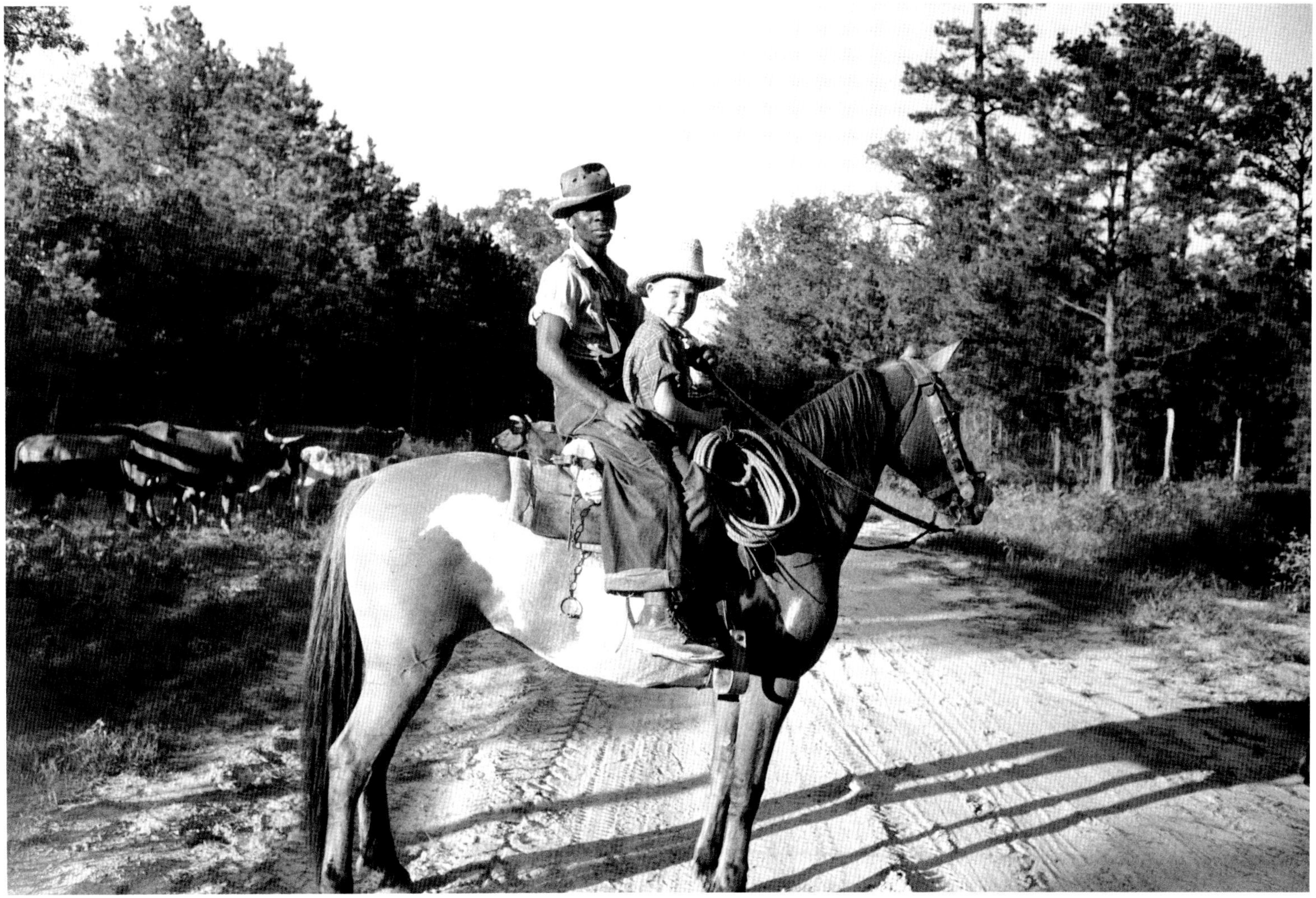

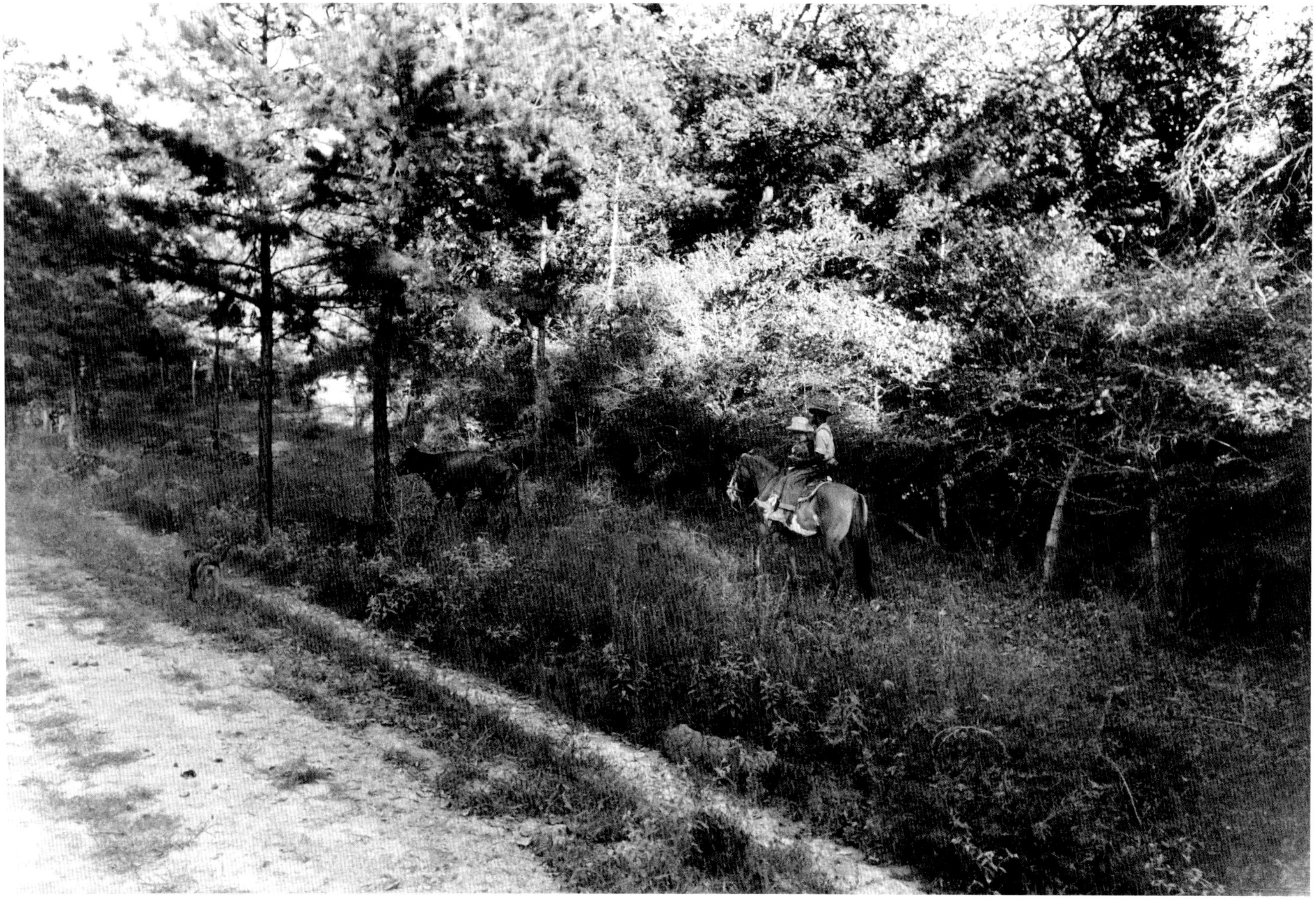

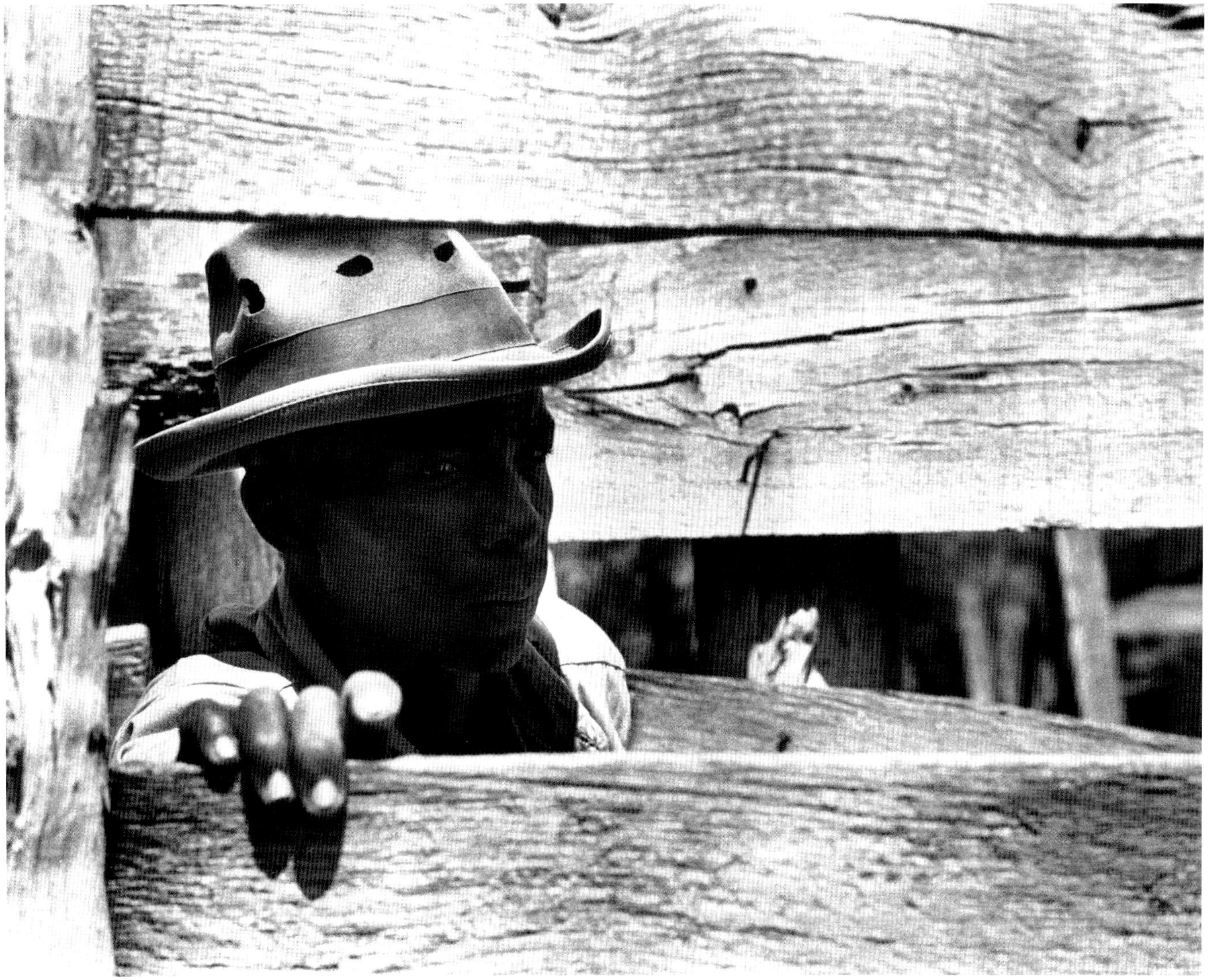

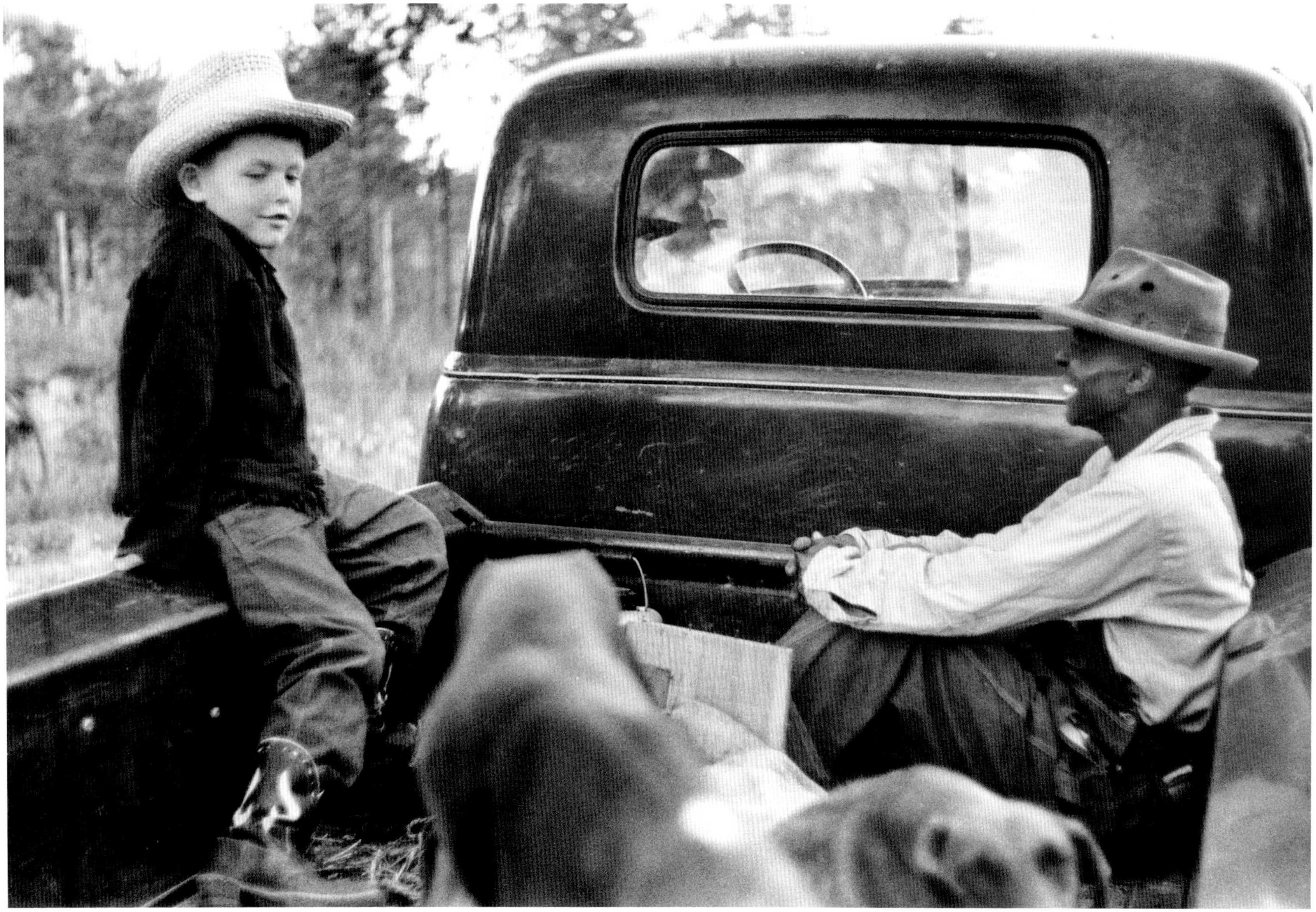

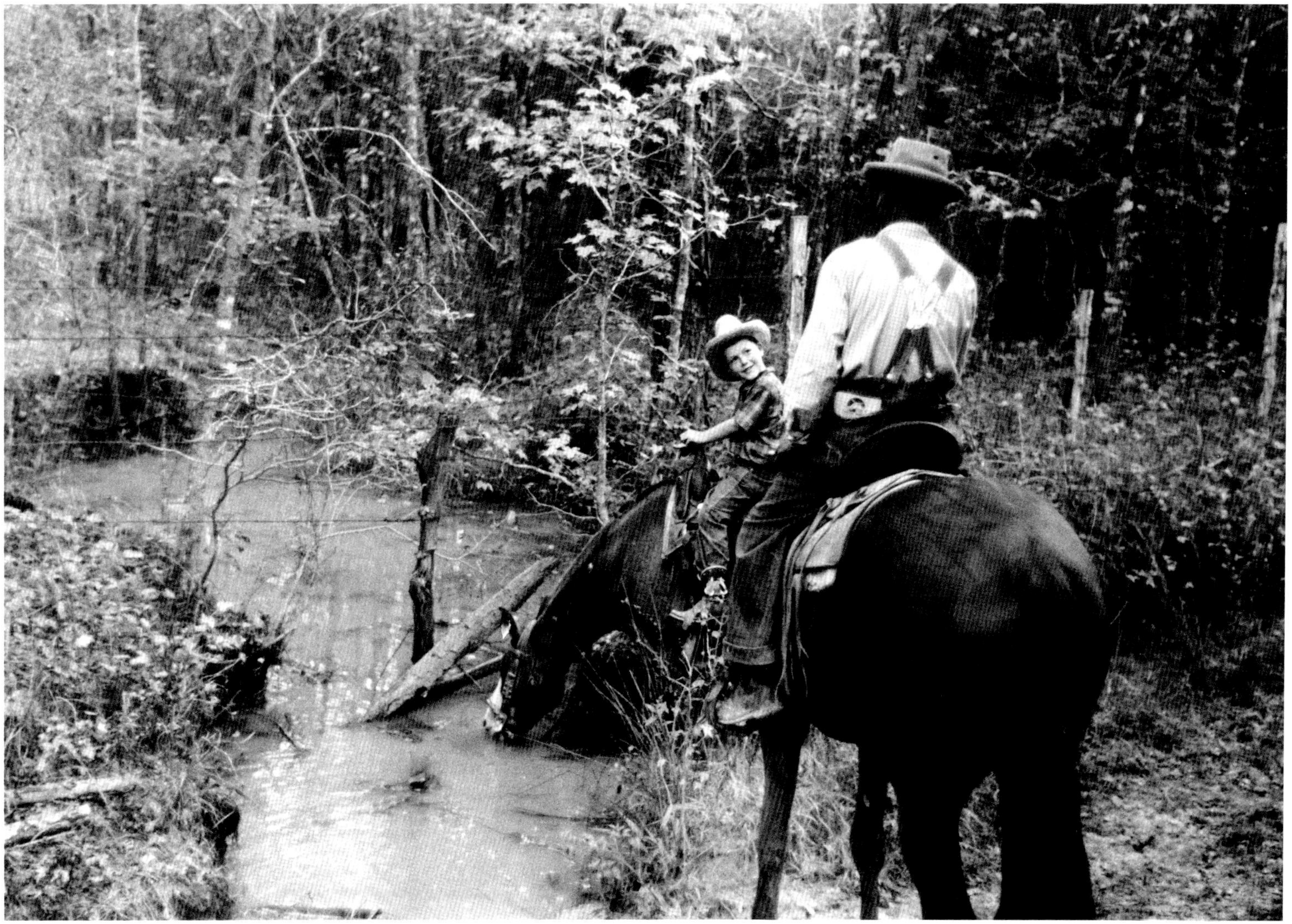

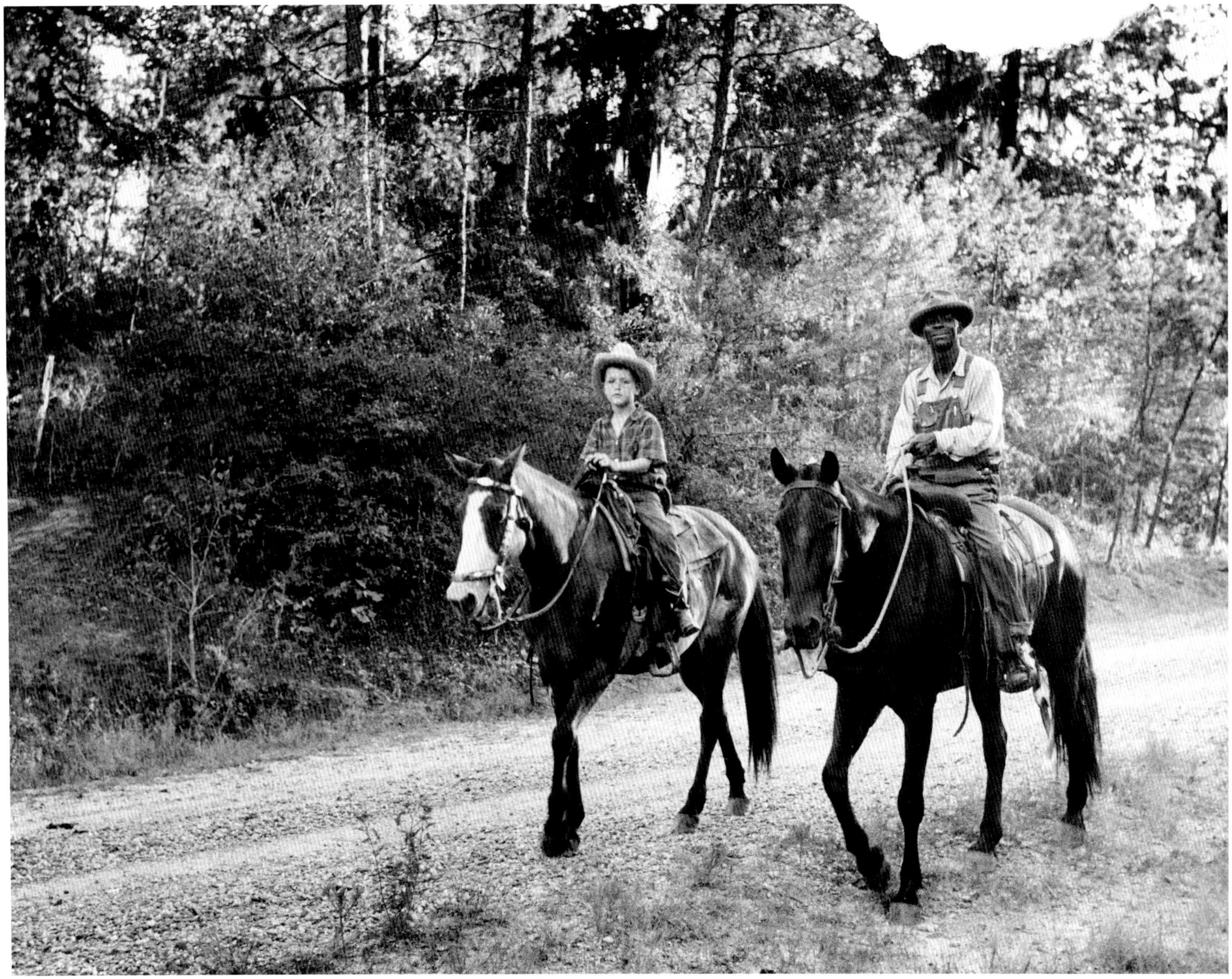

Working the Pens

Cattle do their own work in the pasture, on the range. Ranchers' and cowpunchers' works revolve in some way around the pens. Sorting, separating, branding, doctoring, shipping, all these works come through here, come through these places not born of the land itself but birthed by the ranchers' hands, the cowpunchers' hands, the hands that cut the posts and timbers from this place, hew them and shape them and assemble them. This space which hands have made is where steps are taken, where the work of becoming is done. Here, calves become cattle, cattle become beeves that will become beef.

Working the pens, we become something, too. For we are not born one thing or another. Not black or white, not even black-and-white, but all those limitless grays. We come into this world a wealth of what could be, a treasury of possibilities. The extent and nature of our future tenses, of our potentials, we may allot those to fate. But only in the sense that the Stoics used the term, that fate operates through our actions, through our efforts, through the works of our hands. We have a hand in our own destiny.

Looking back, we will see in the shaping of our destinies hands other than our own. We have mentors who work with us in the pens. When we are too small, too inexperienced for our own saddle, our

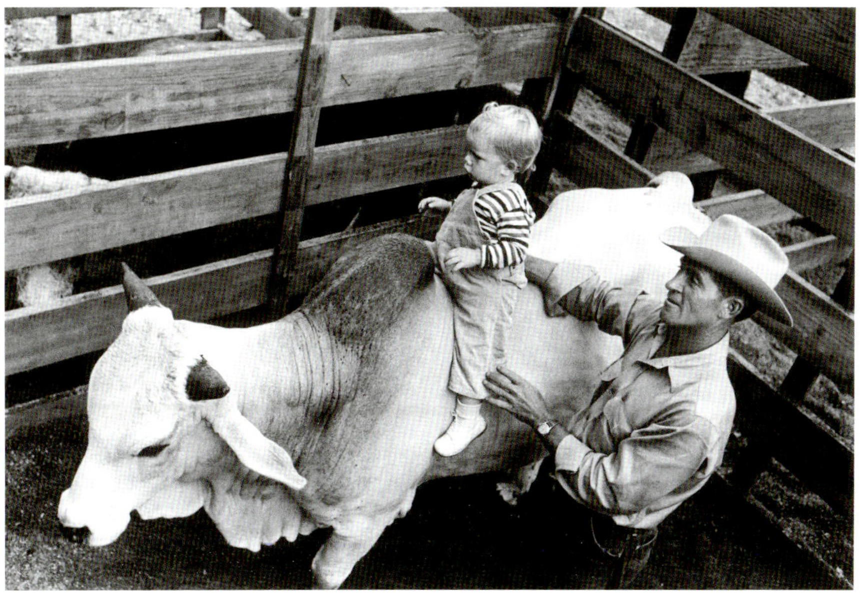

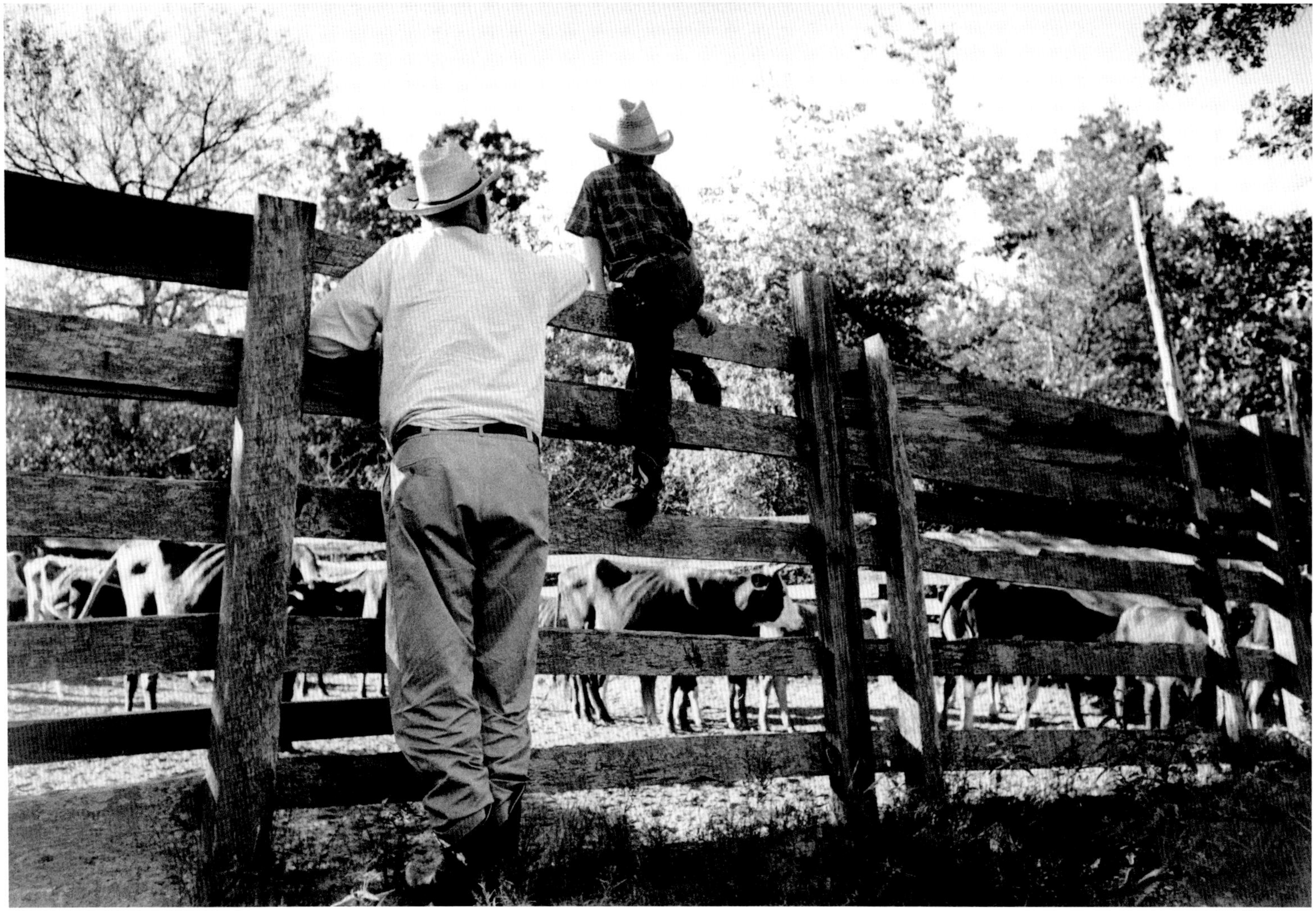

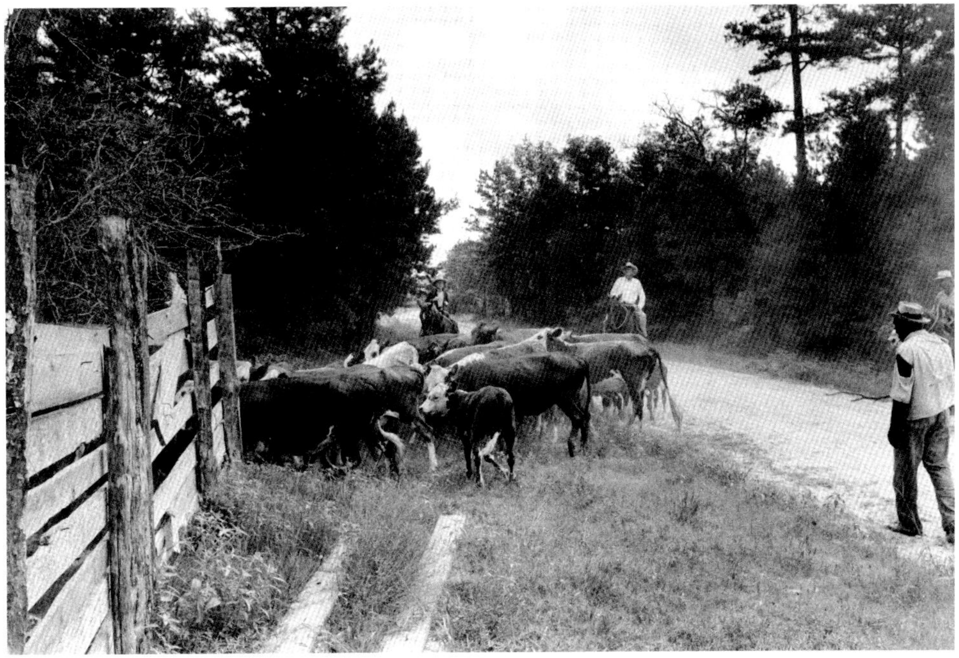

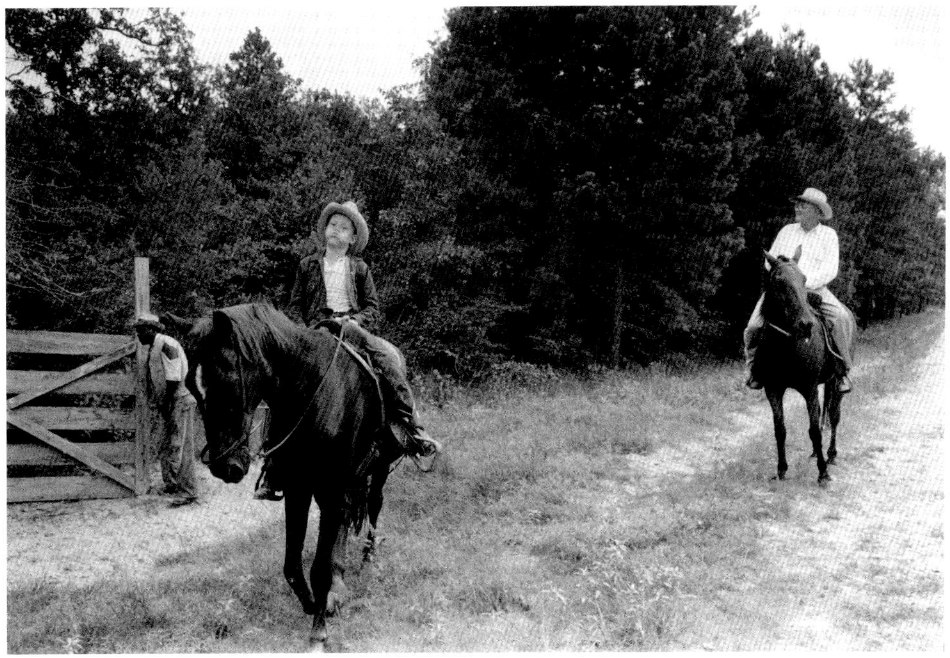

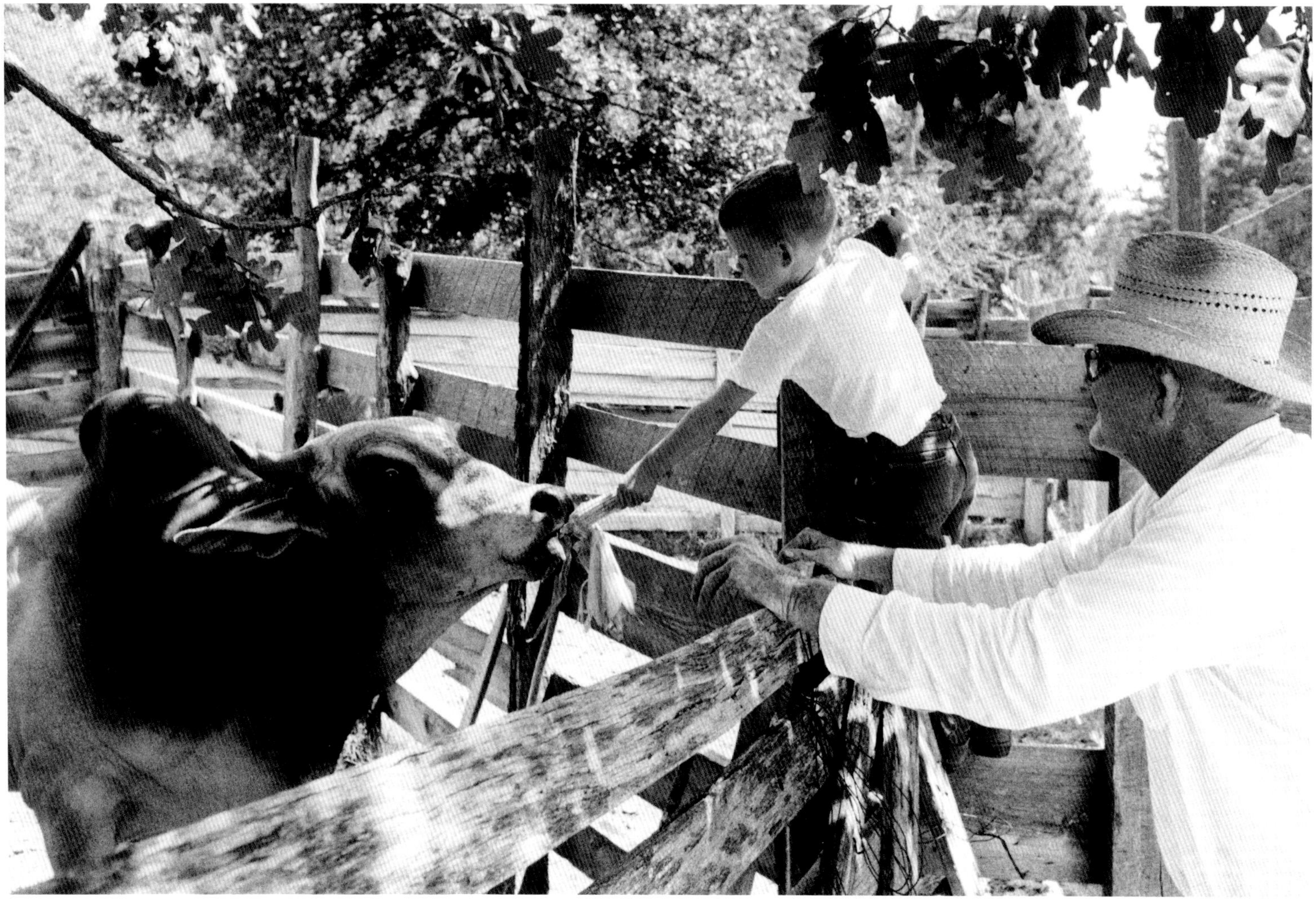

own mount, Dayton, the cowboy, puts us first in the seat with him. As we grow, he rides alongside us. He laughs with us, sometimes at us. He gets out of the pickup or down off the saddle and catches the gate. He shows us how to be cowpunchers without telling us. We are not his equal, his colleague, but we will become that.

Looking back, we will also see and feel the hands of the builders who are more than mentors. Who work on us more than with us. Builders whose hands are of the same flesh as us, flesh that springs as much from the soil and the grass and the water of our place as it does from the genetics of our tribe, our family. Hoyt Porter is of this land, and we are of him. We are not his equal, his colleague, but we will become him.

This kind of mentorship, this kind of building, this kind of working the pens we call love. It permeates these photographs just as thoroughly as the light that saturates them.

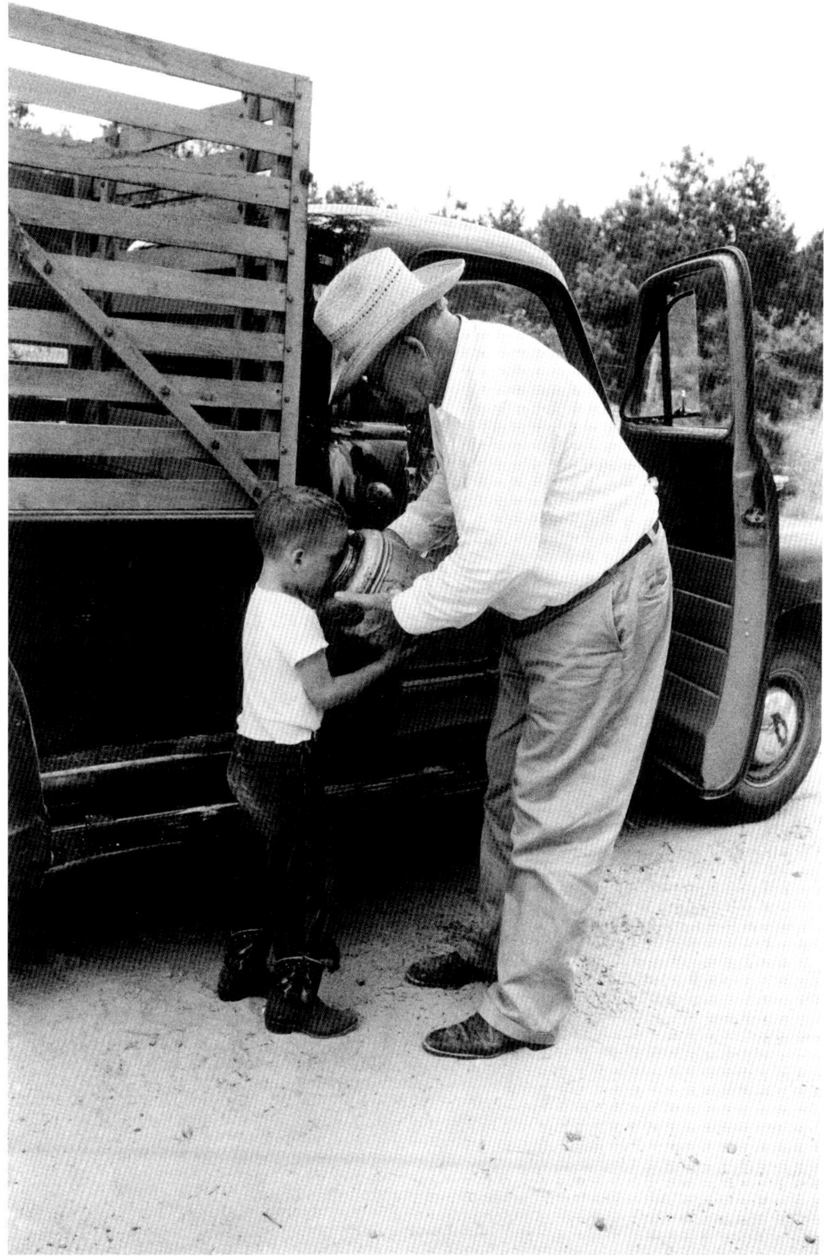

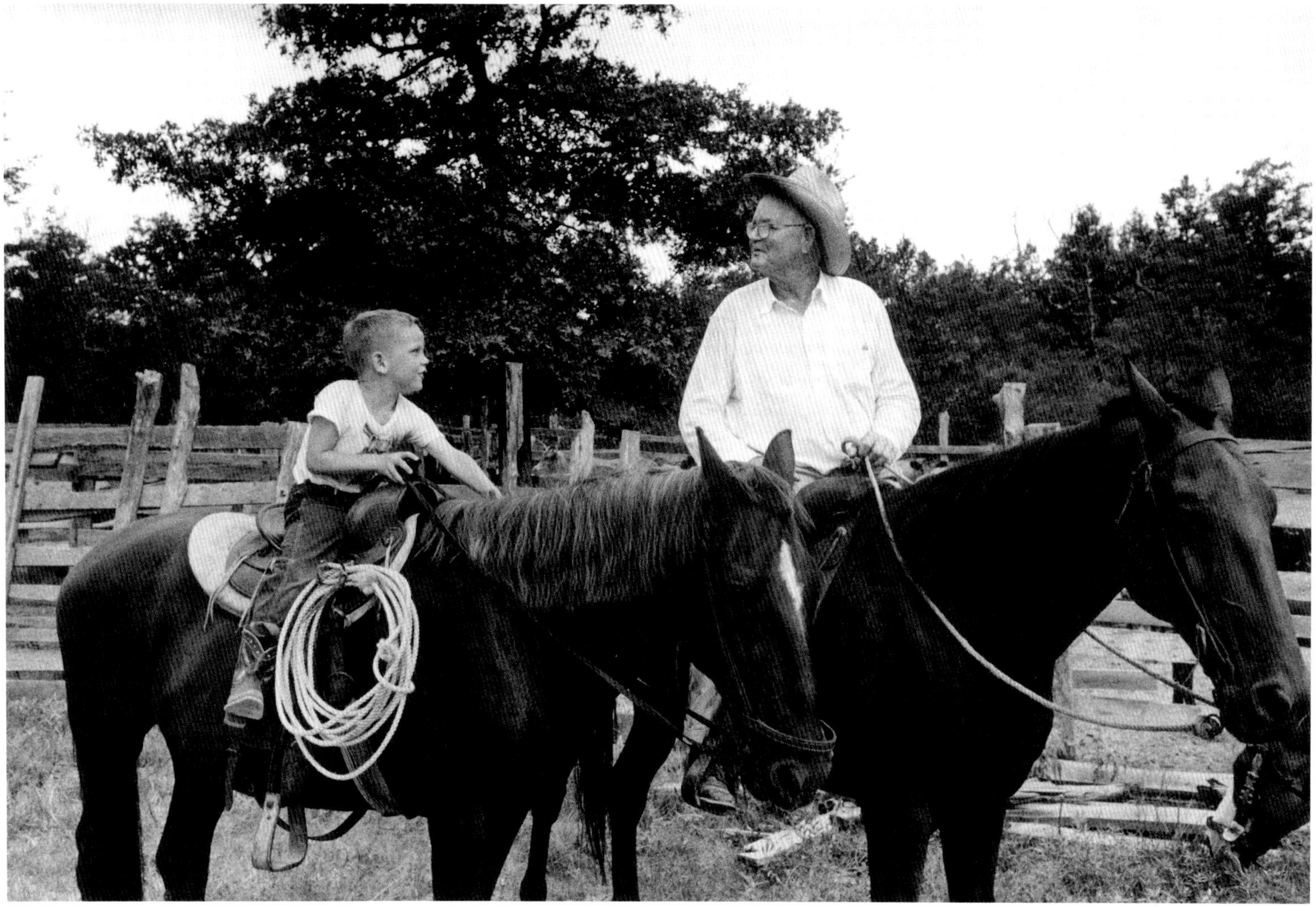

Dominos at the Garage

Midday in the summer and it's no time to be out in pasture or field. Better the service station where there's shade and a fan and a table and friends enough for a game of Forty-Two. Might even be a young woman in a bathing suit stopping in to get gasoline, maybe not stopping the game but certainly stopping the conversation.

Another day, the grandsons take turns watching. The slow shuffle scraping across tabletop, the dominos' muted clicking and clacking, each player arranging the draw to their own taste, some holding them like cards, some standing them up on edge in a half-circle. One generation passing it down to another. That's how you learn Forty-Two, a game birthed in Texas in the late 1880s but for which there were no published rules until over a century later.

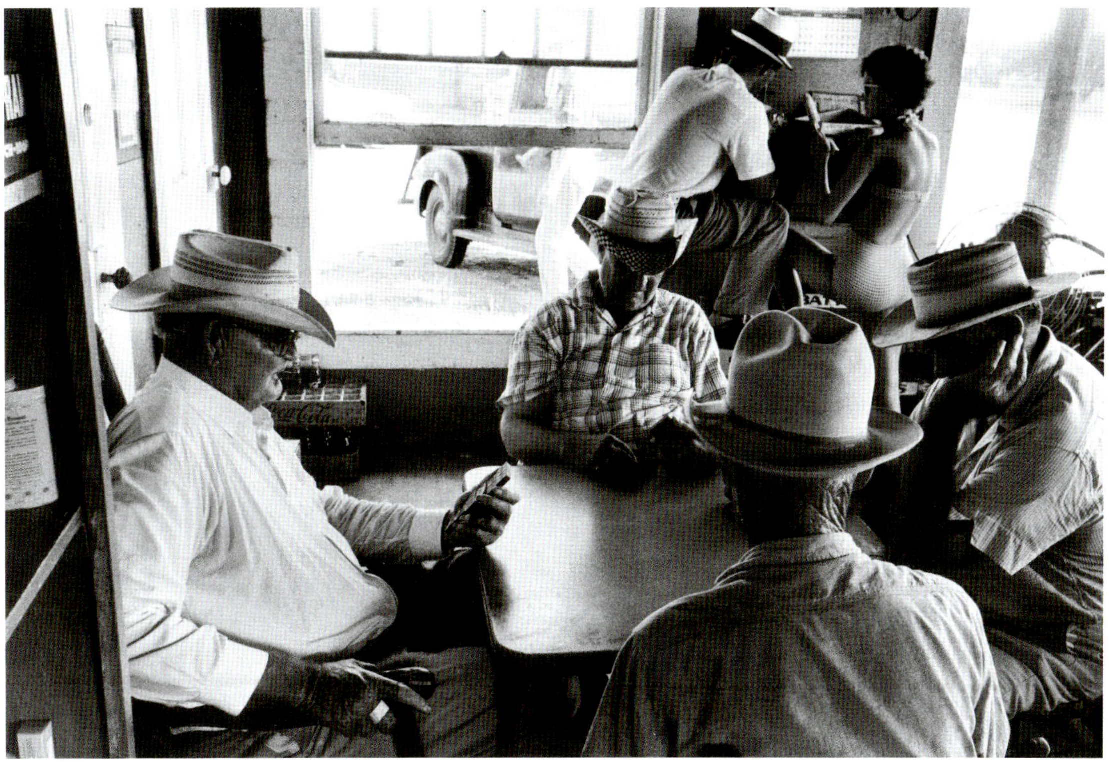

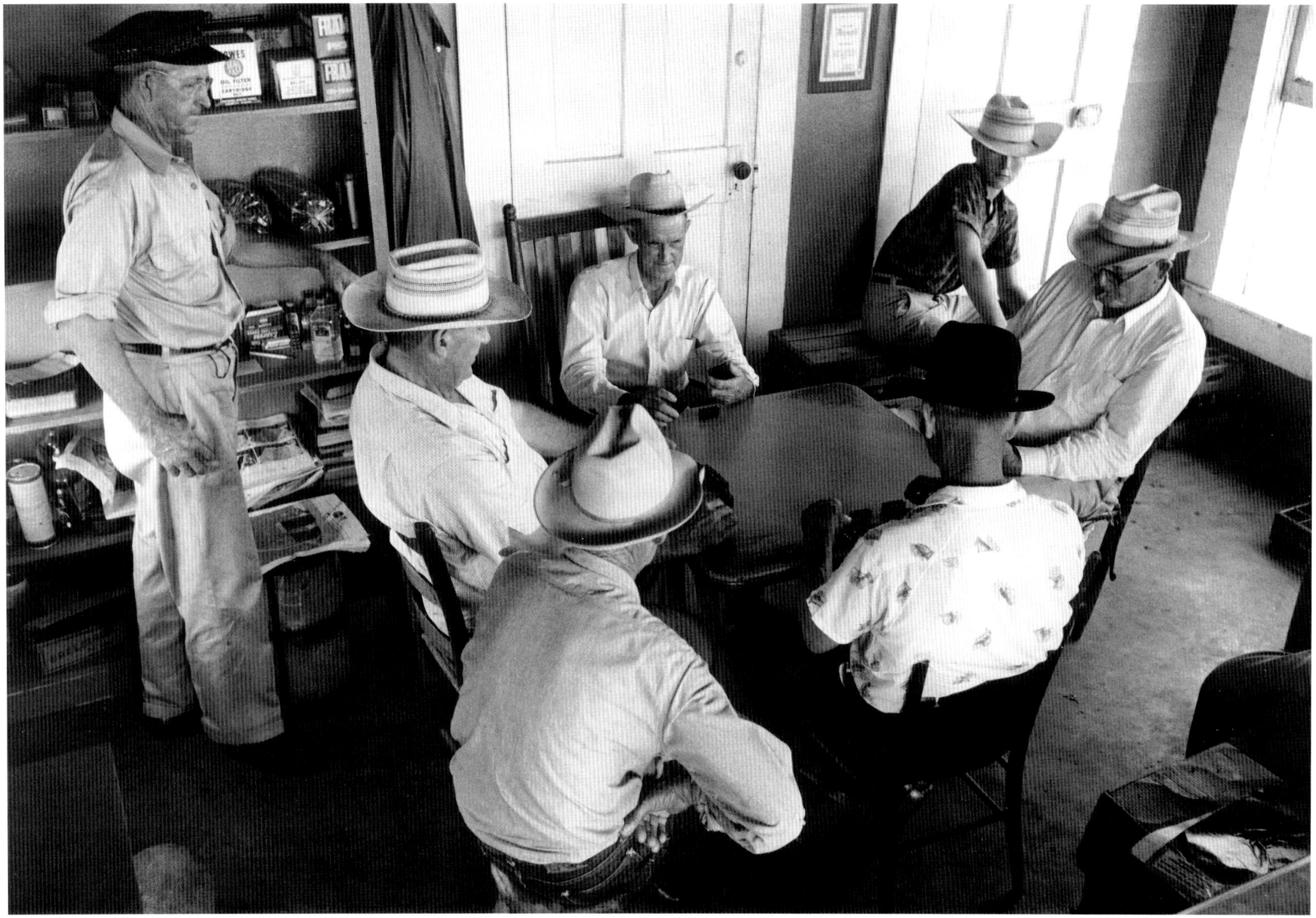
OIL FILTER
CARTRIDGE

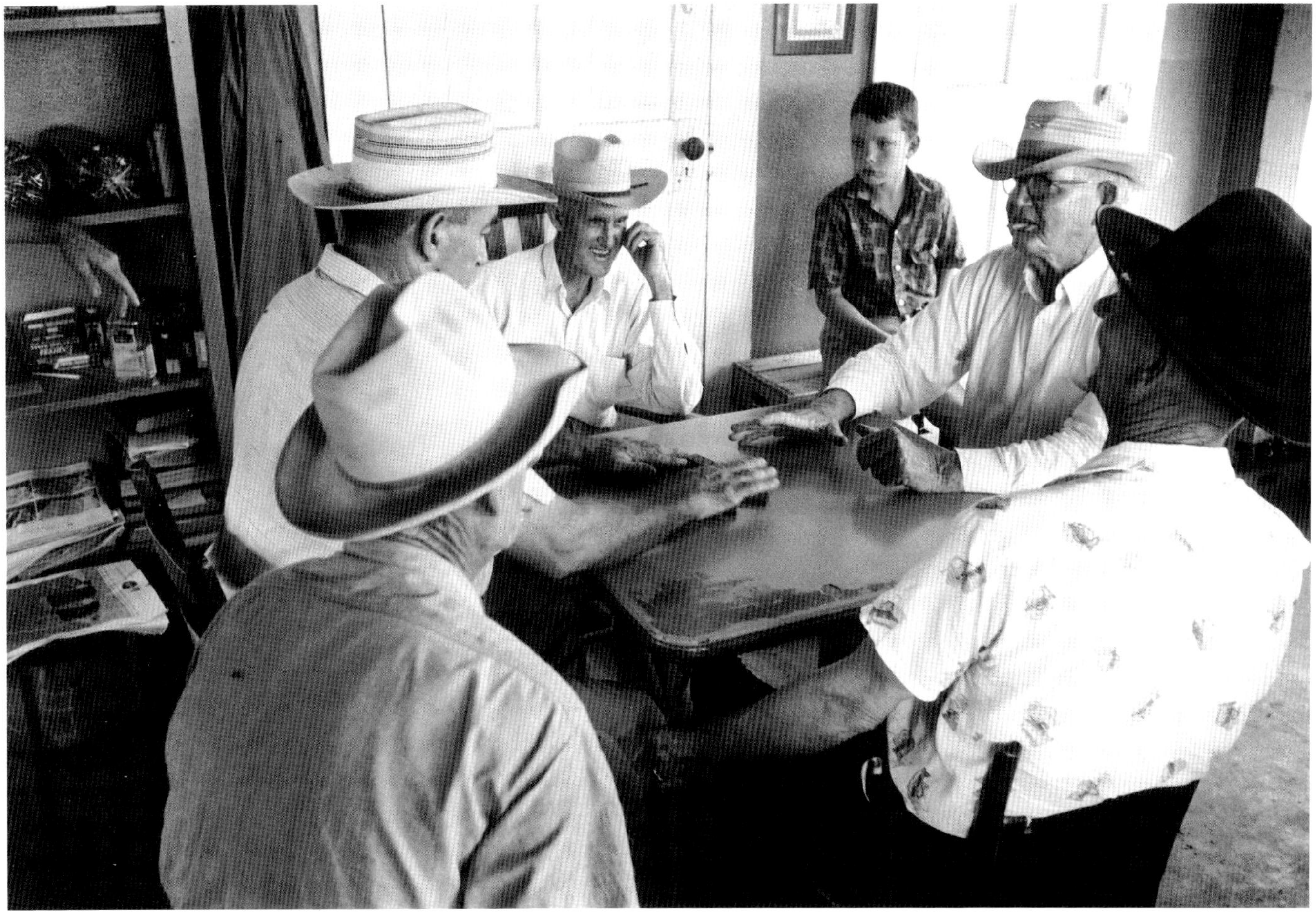

A bidding game as much strategy as luck, it moves at its own pace. It can be slow, or quick. A player has to think a lot about what the other players have got and let on very little about what they've got their own selves. When the game's slow and it's not your move, you can even nod off.

One December day in 1974, Hoyt drifted away during a game. His turn came, but he never woke up.

Church Work

Reuse and recycle. It's a modern mantra, in some circles today a political statement, an avowal of righteousness, a way of looking at the world. Not here. Not then. In Lovelady in the 1950s it's a way of life. It's how things are done, so much a part of the ordinary, the everyday, that folks would think it odd to single out, to make fuss over.

The church needs work, repairs need to be made. In the next town south there's an old house, vacant, ready to fall down. Like all things neglected in a community, an eyesore, even dangerous. Yet, like all things neglected, it still has worth. In some ways more valuable, the wood seasoned by weather and long, good use.

On the horizon, the smokestack at the sawmill belches out waste. But here, Hoyt Porter's fellow congregants mend — as best they can — the damage done by such squandering. With nail bars and crow bars and hammer and sledge, with muscle and sweat and humor and fellowship, they dismantle what is no longer useful so as to renew what yet remains of use.

Pieces of watermelon speared on a pocketknife blade, a Barlow, maybe, or a Case, cantaloupe eaten from the bowl of its own rind, black and white and farmer and cowhand take a break together. In Guy's photograph, a black arm and hand leads your eye to a white arm and hand, the white hand holding up a timber, the timber becoming the pillar of a cross.

The work of the church made evident.

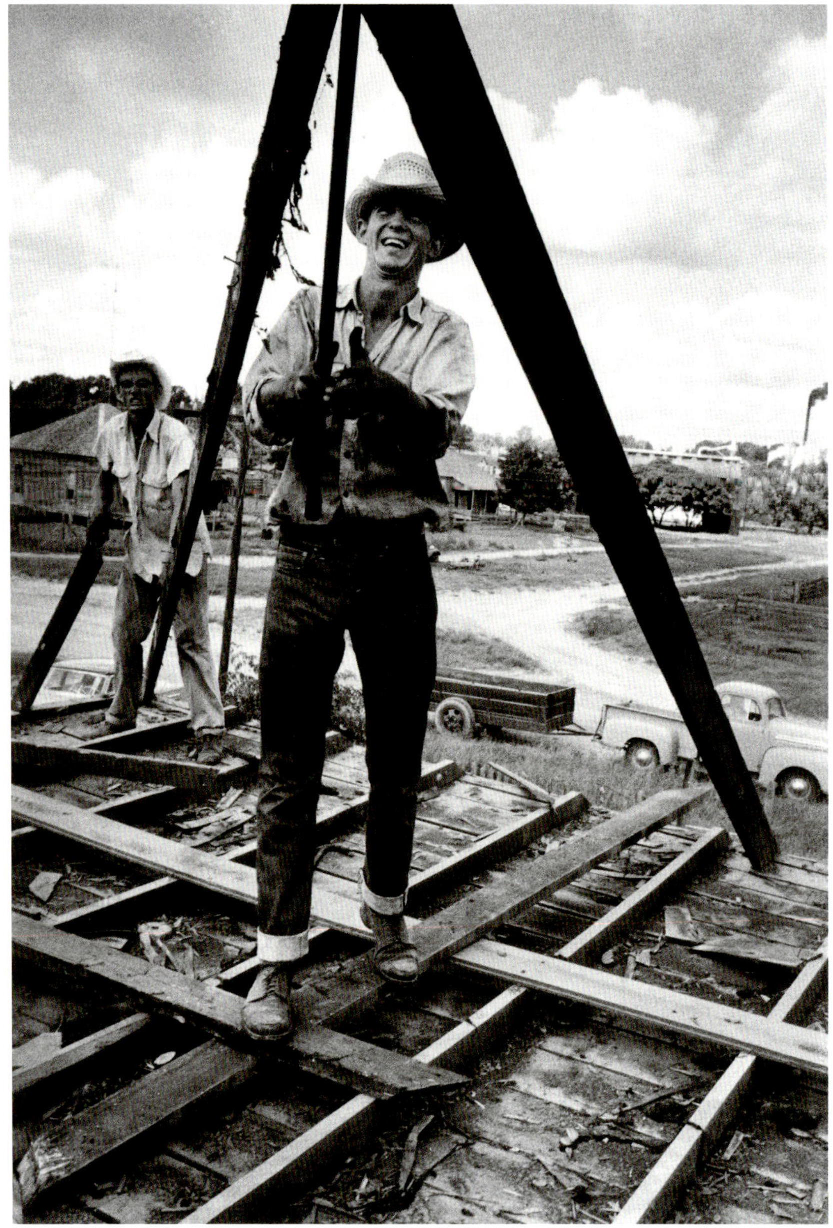

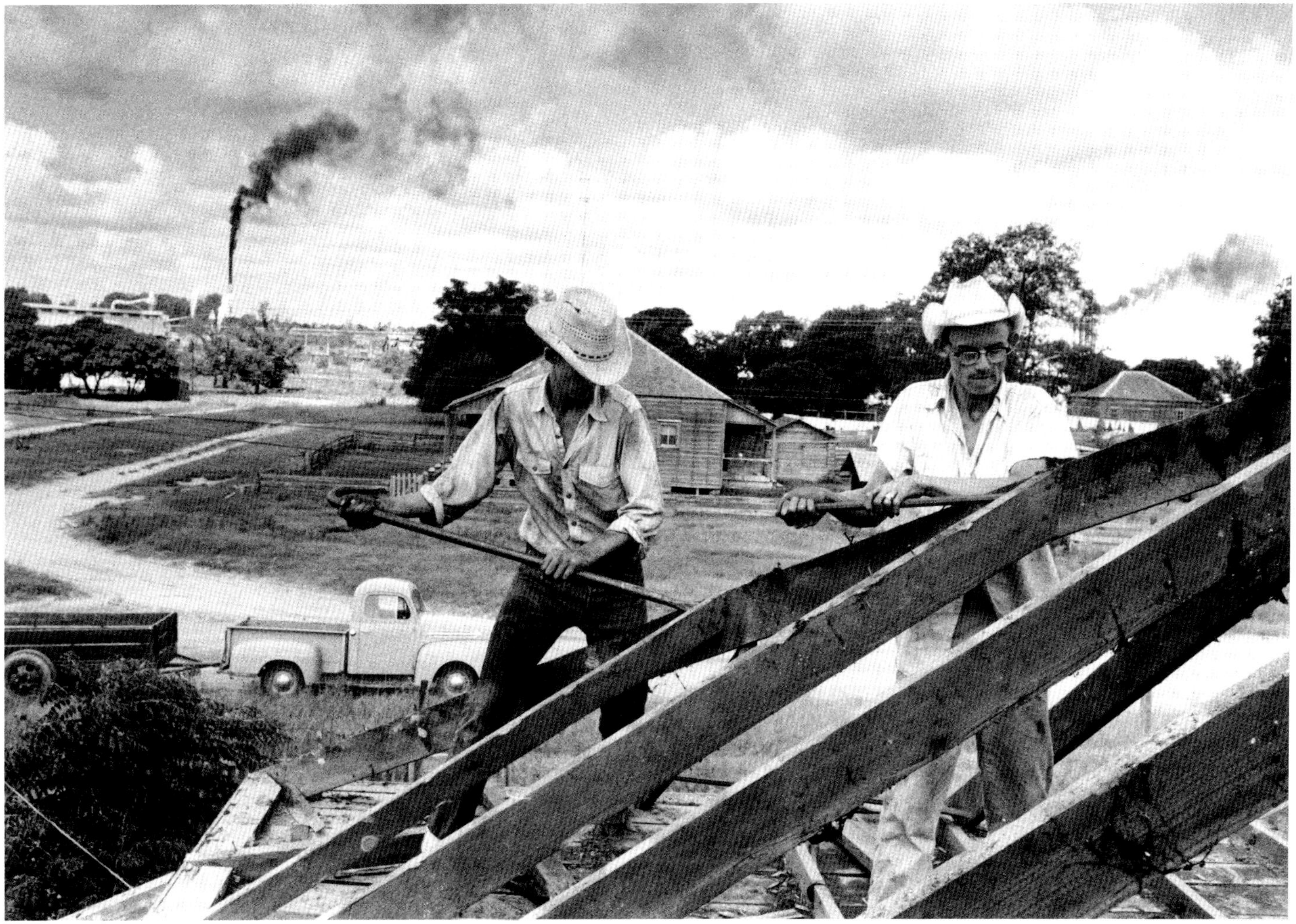

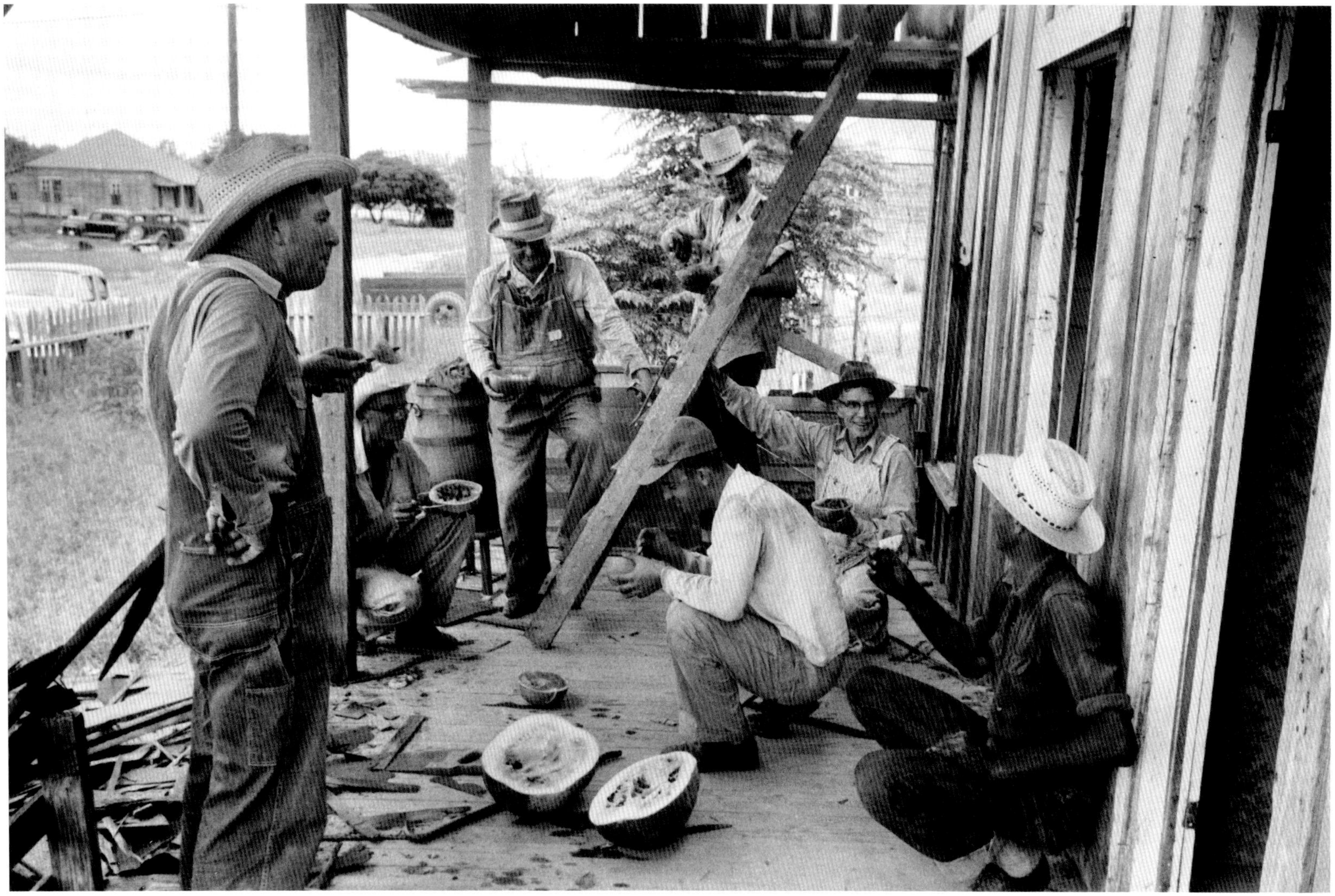

Antioch Homecoming, July 1955

Not the time of year for crisp fall afternoons and corsages made of mums and school color ribbons, not the time for the glitter and sparkle of marching bands, not the time for ex-students in the stands. The alumni never leave here. It is the students who return.

The first grave site was set in this place in 1870, when it was called Neville's Prairie. The town got a post office and a new name in 1890; both were gone by the end of the 1930s. But Antioch Cemetery endures, refreshed each year by those who come home.

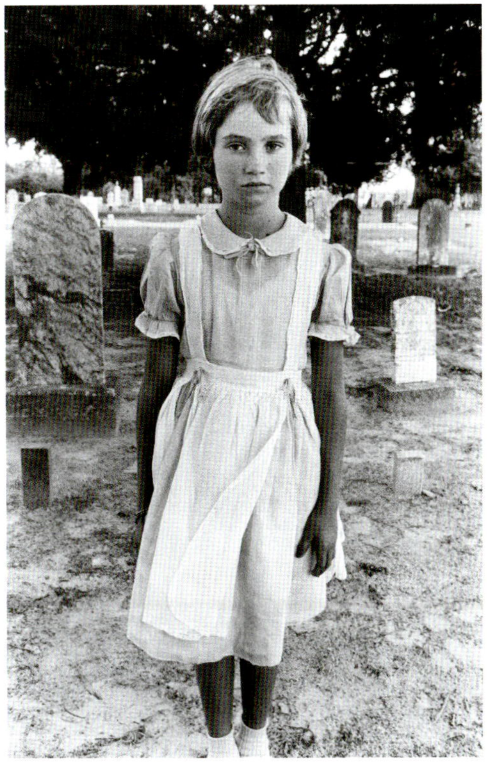

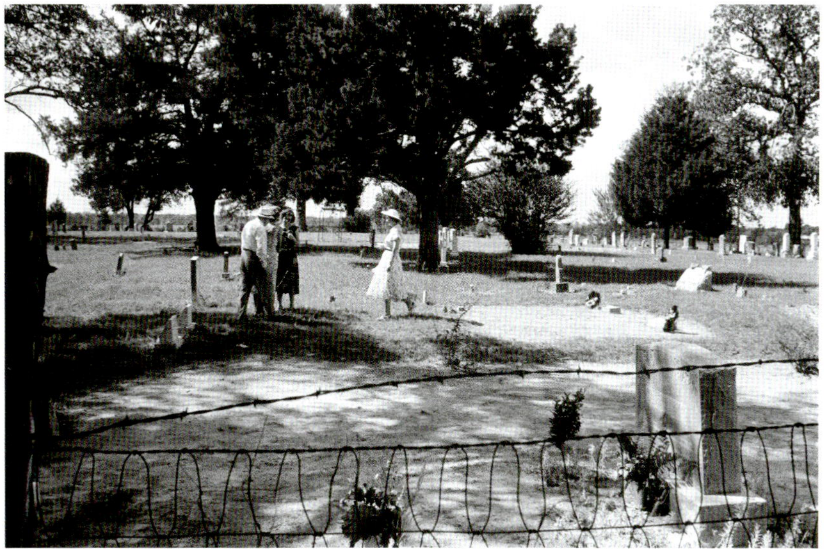

Guy comes down from New York with Doris and the boys. He brings his camera, as always. In the City last year he took Audrey Hepburn's portrait. Here, he takes the portrait of a community. A student of the human drama, his scholarship is in his images.

He is not the only student. Others come to build long sawhorse tables in the shade of the oaks, covering boards with cloth and loading them with potluck casseroles and fried chicken and mashed potatoes and jugs of sweet iced tea, loading them with pies and cakes and cookies and hand-cranked ice cream. City folk would call it lunch, but in the country it is dinner on the ground, not a quick bite in a short break in the middle of a day-planner, but a time to take time.

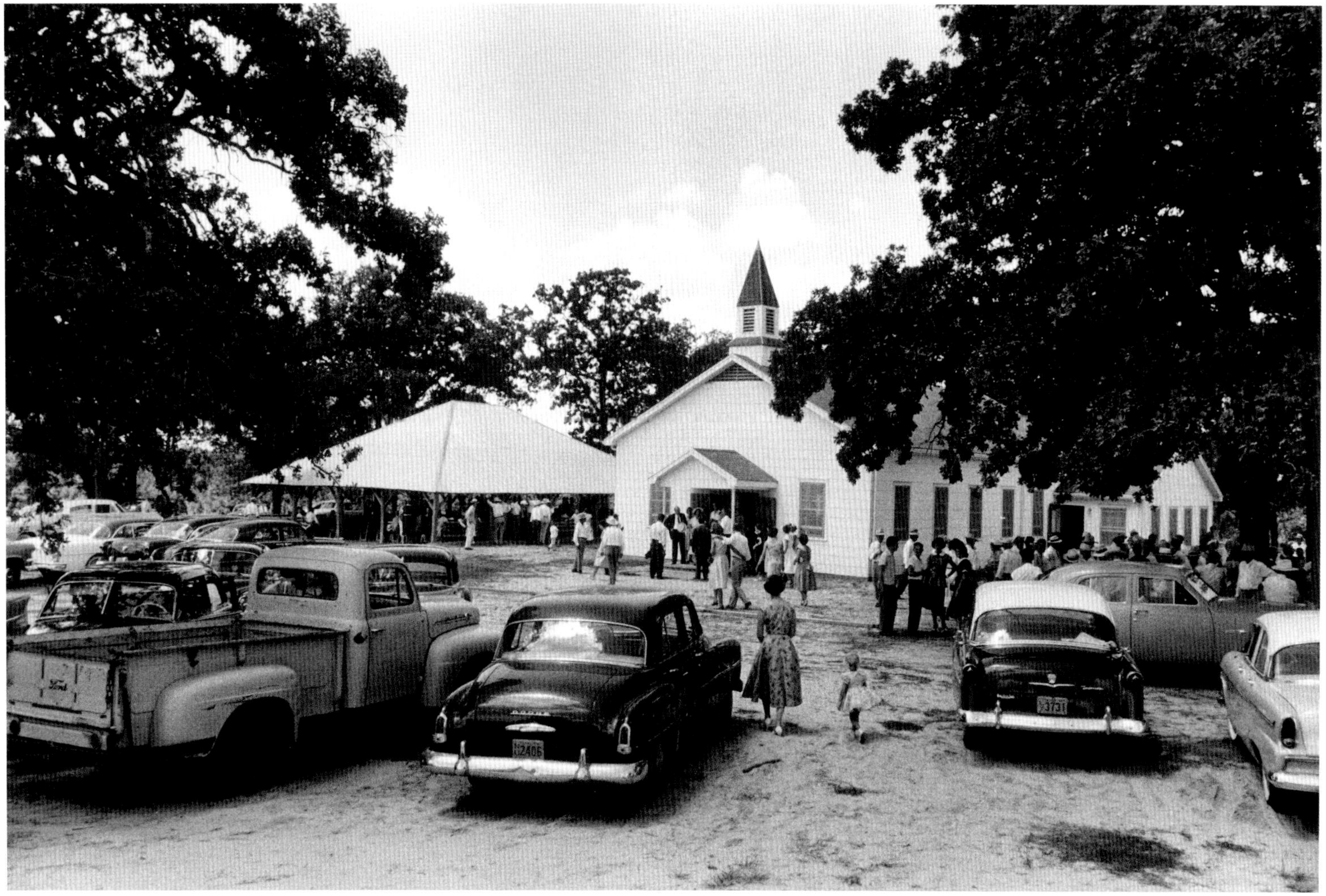

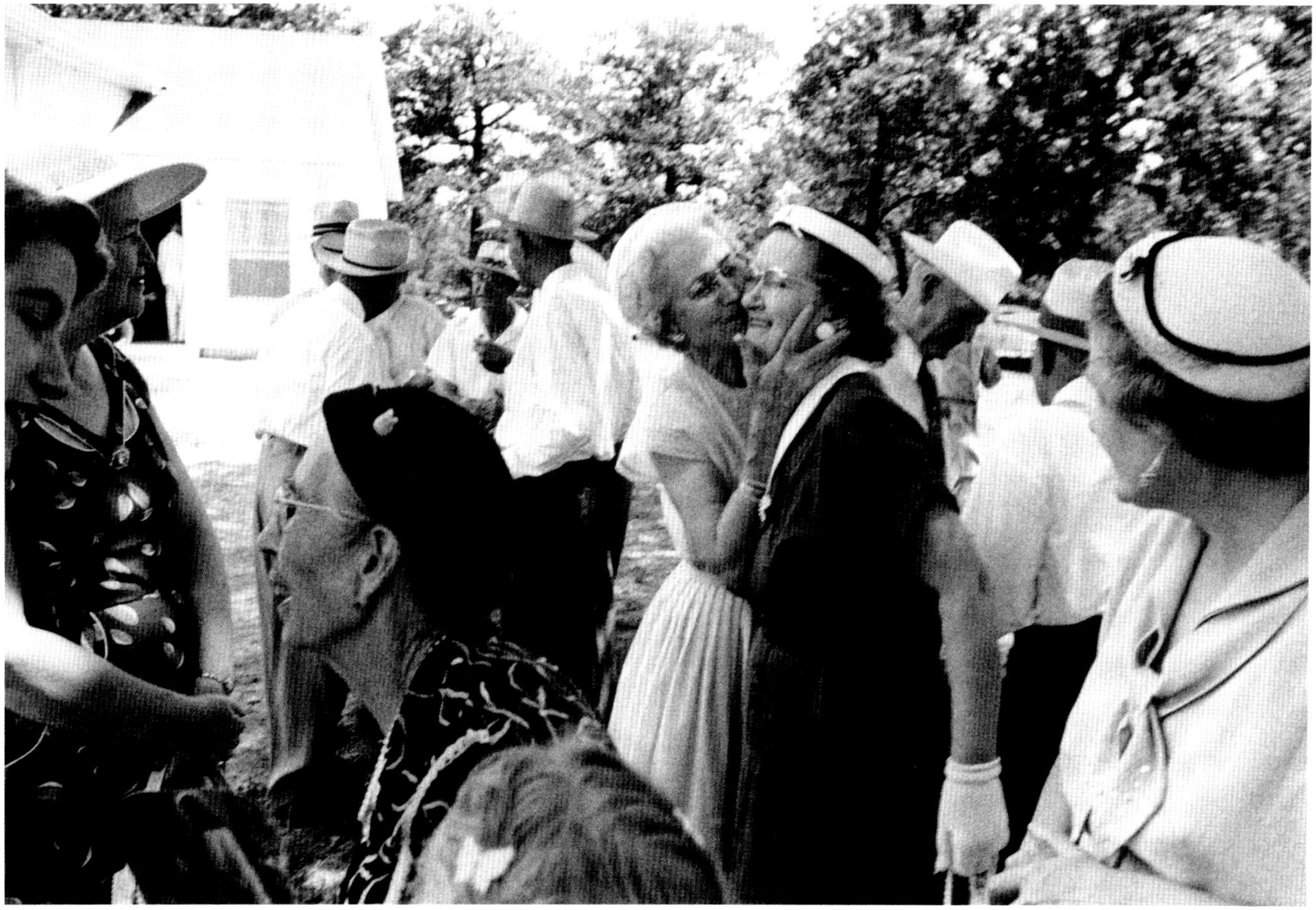

Time to look over the home-cooked dishes spread along the tables, time to go back for seconds, time to catch up on old friendships, time to show off new babies. Time for a conversation, time to visit the family plot.

After dinner, there is singing in the open-air tabernacle — four-part harmonies in "Shall We Gather at the River?" and "The Old Rugged Cross" and "Leaning on the Everlasting Arms." They finish with a business meeting in the Antioch Baptist Church. There is a year's worth of work to be organized: grass will need to be cut, weeds and brush will need to be pushed back, stones will need to be straightened. There is next year's homecoming to plan.

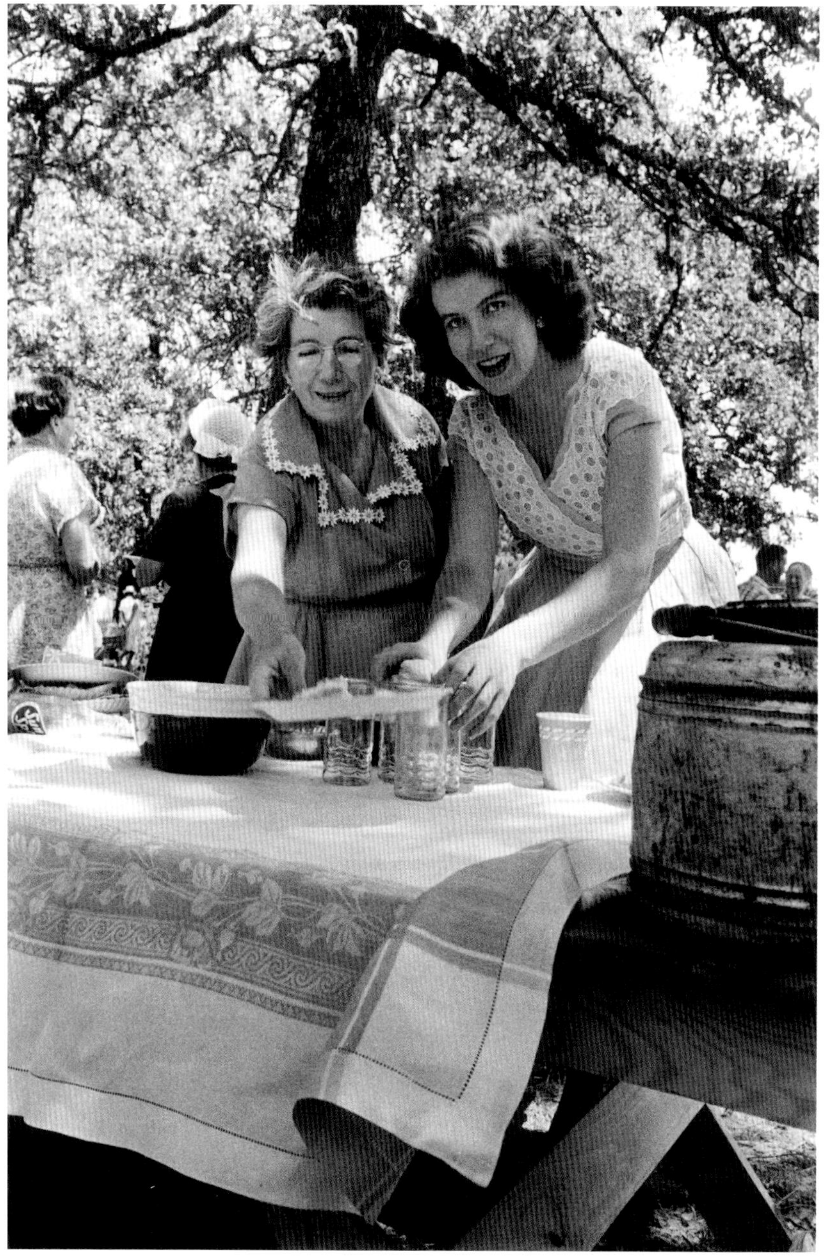

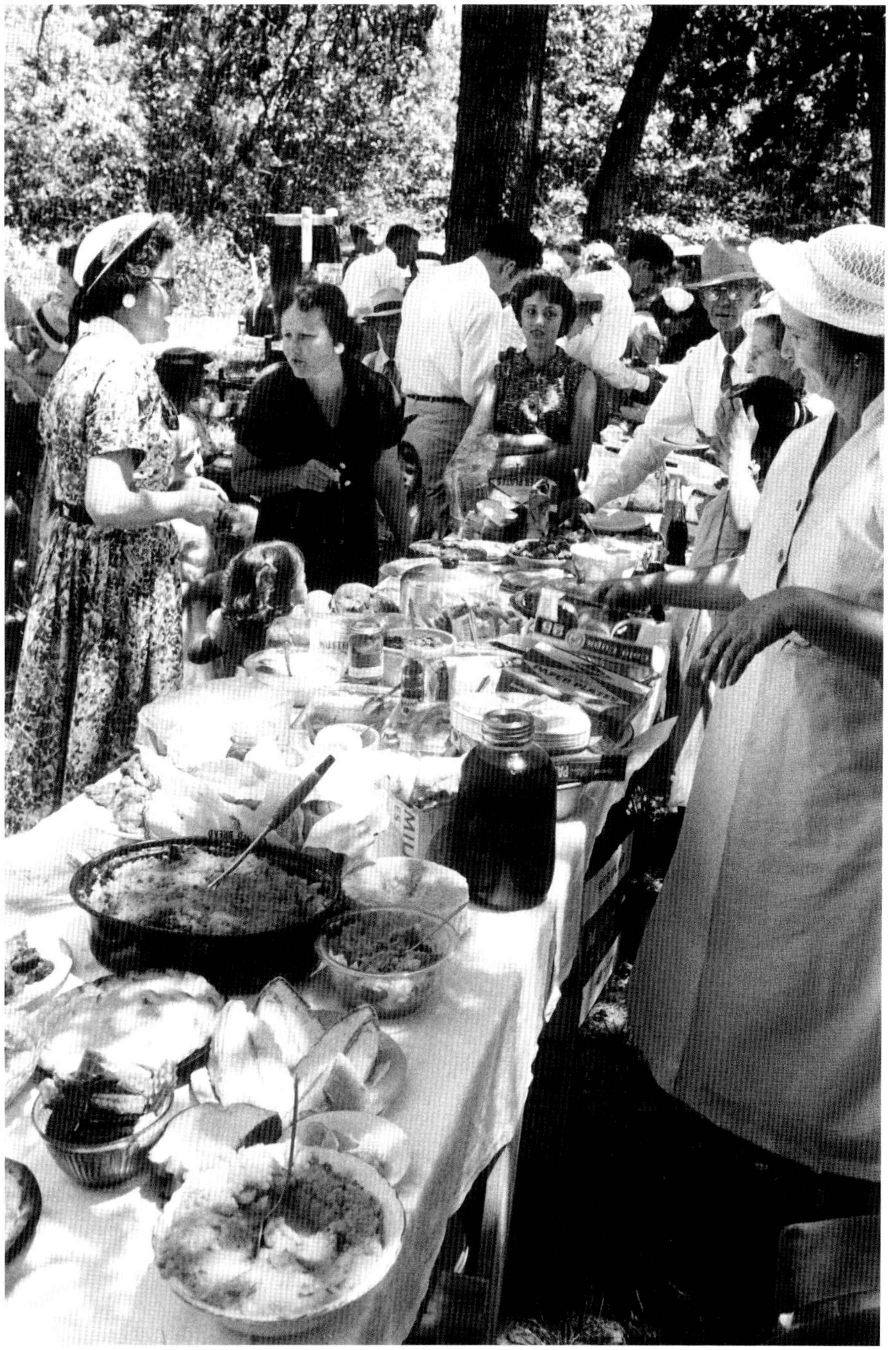

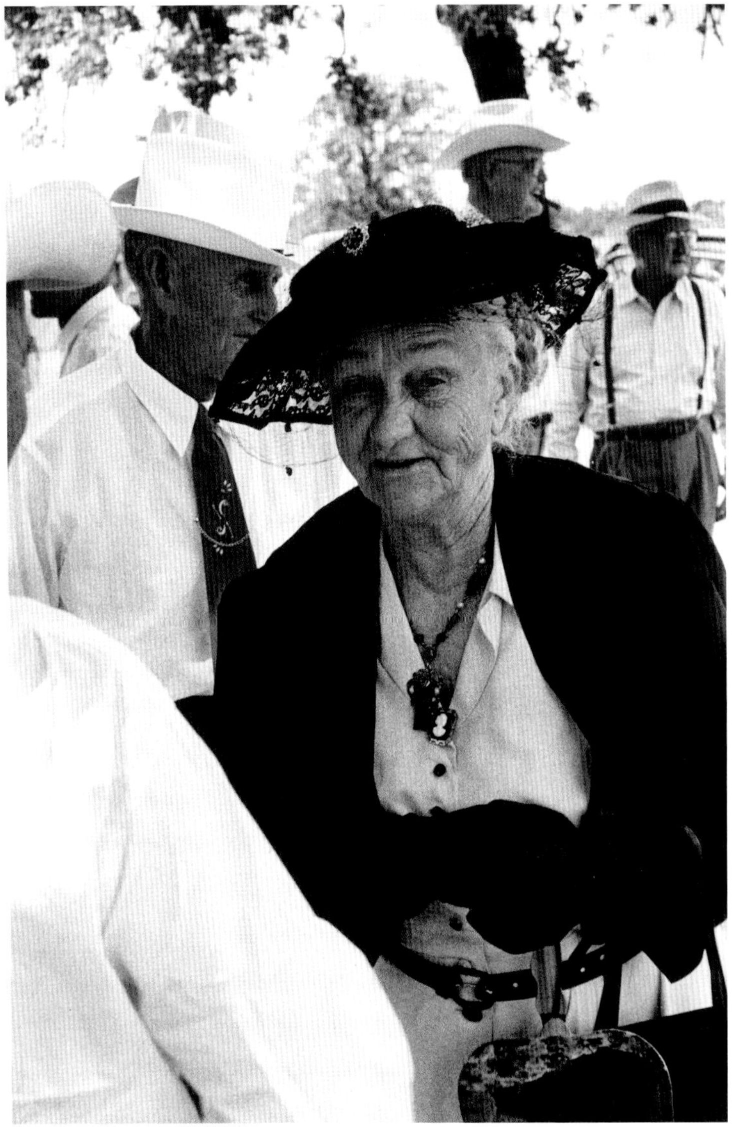

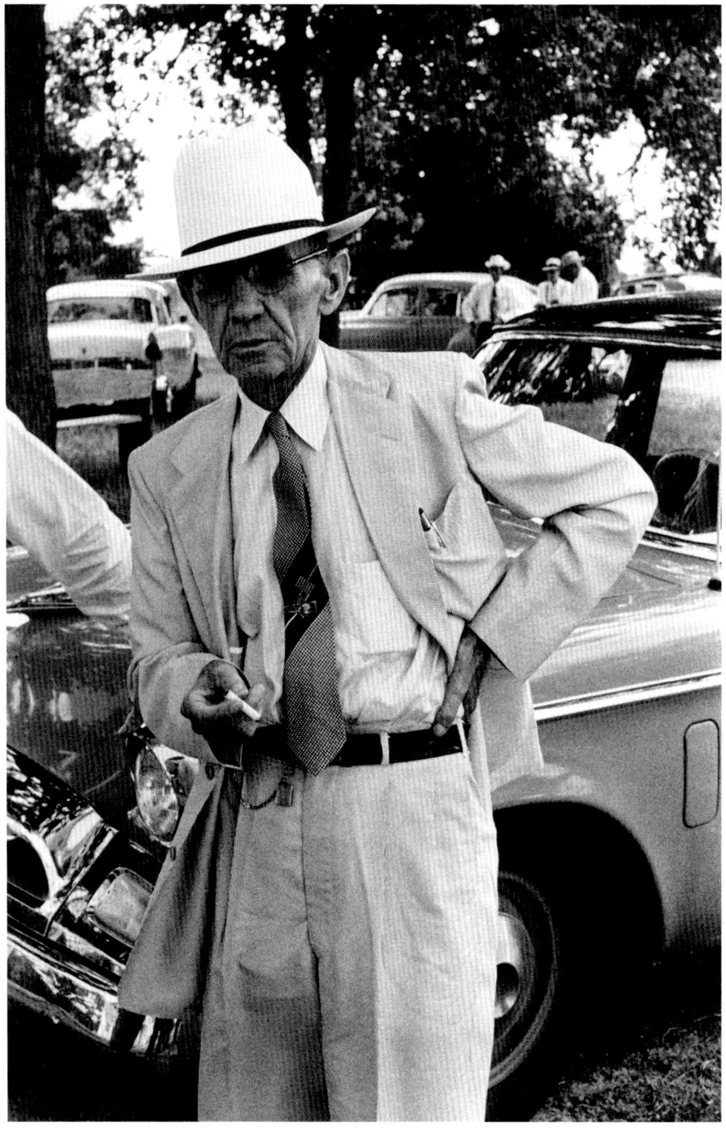

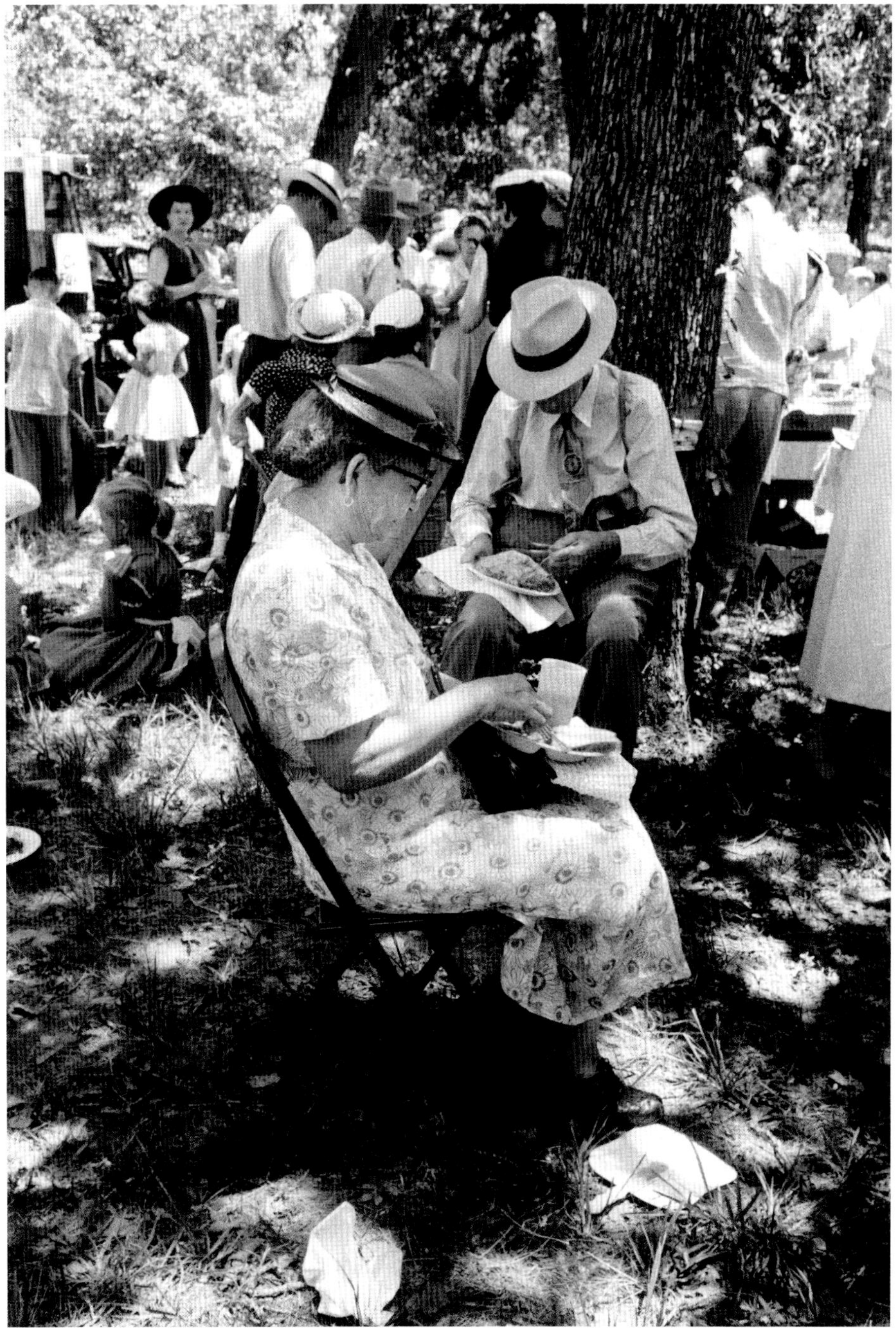

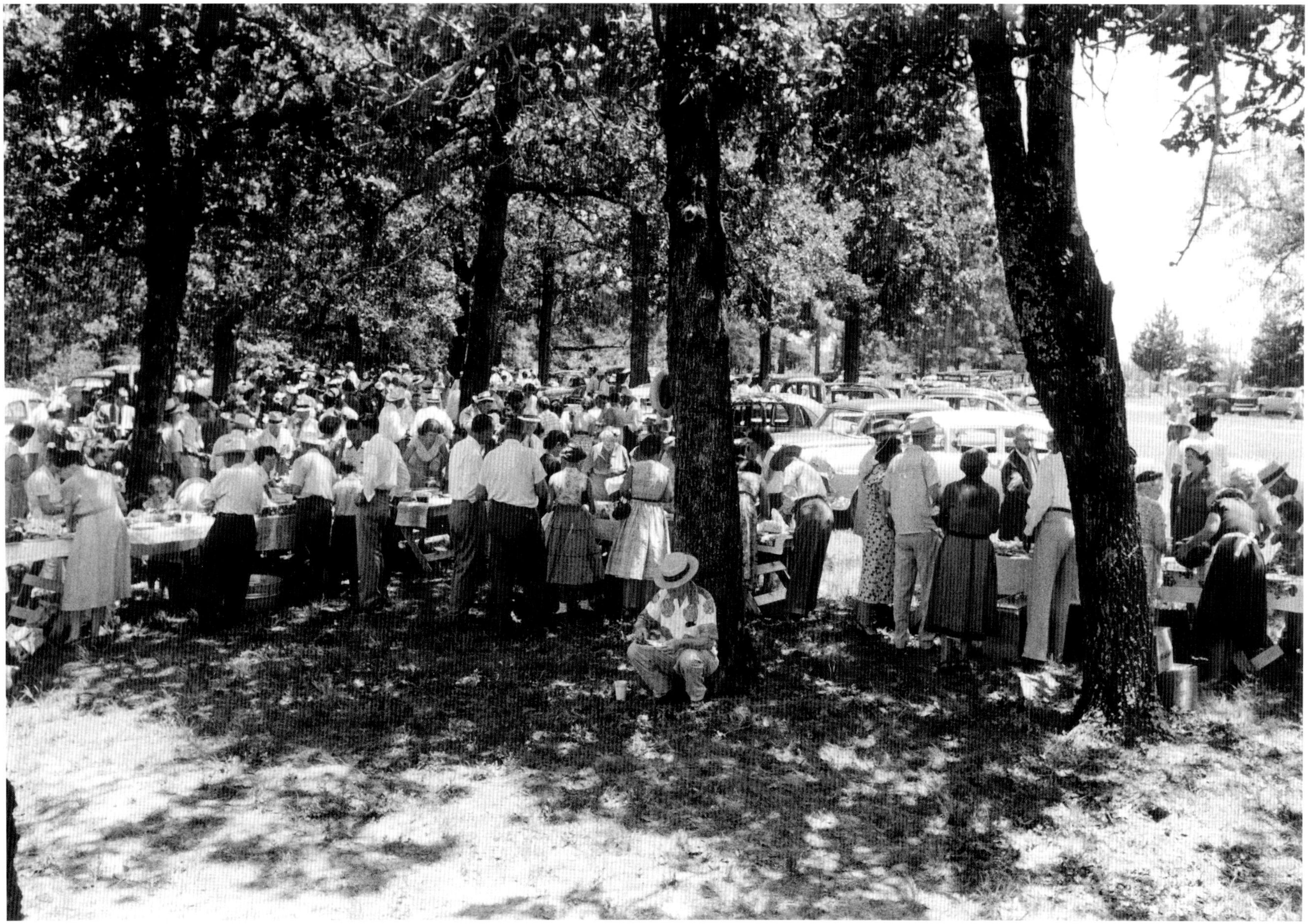

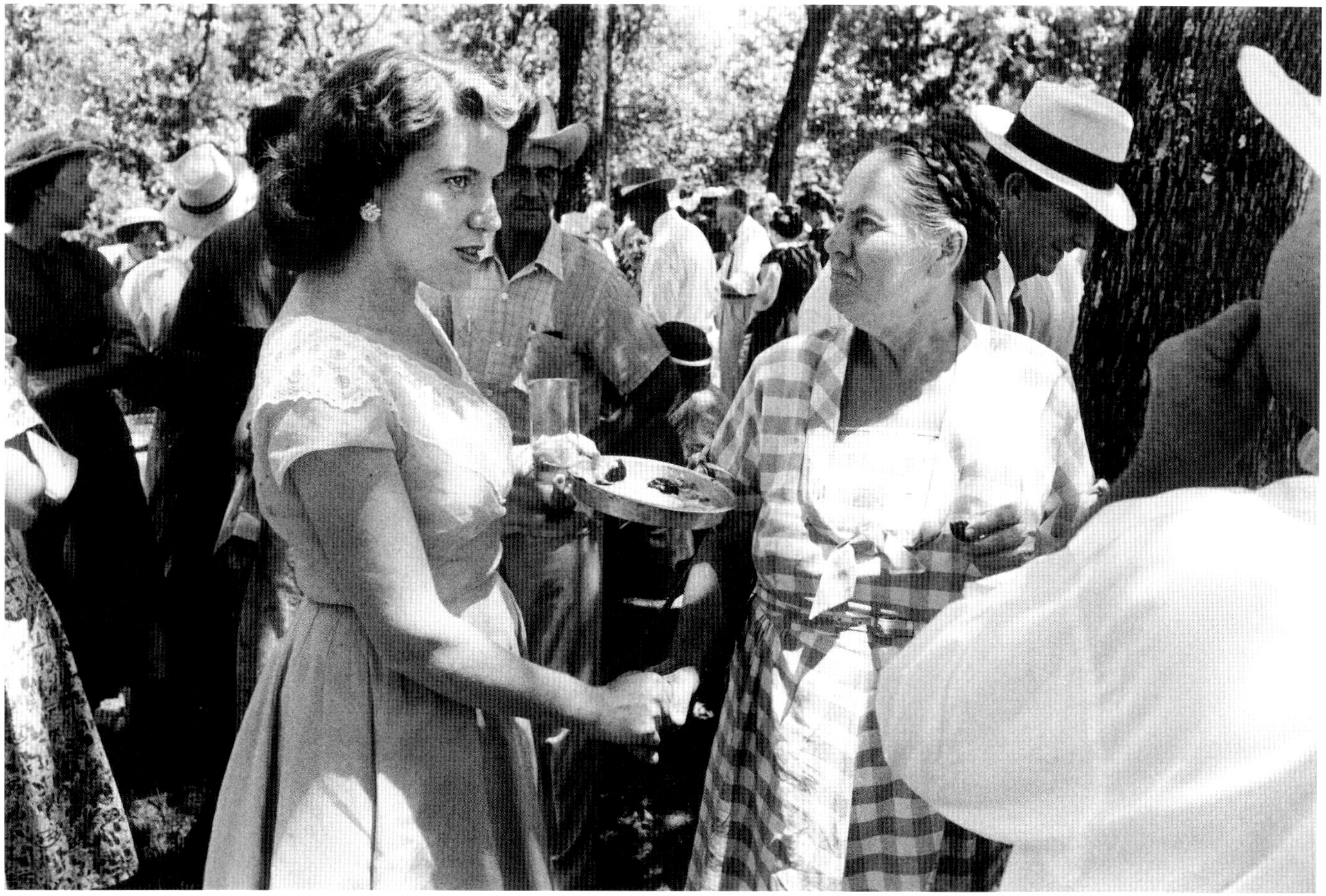

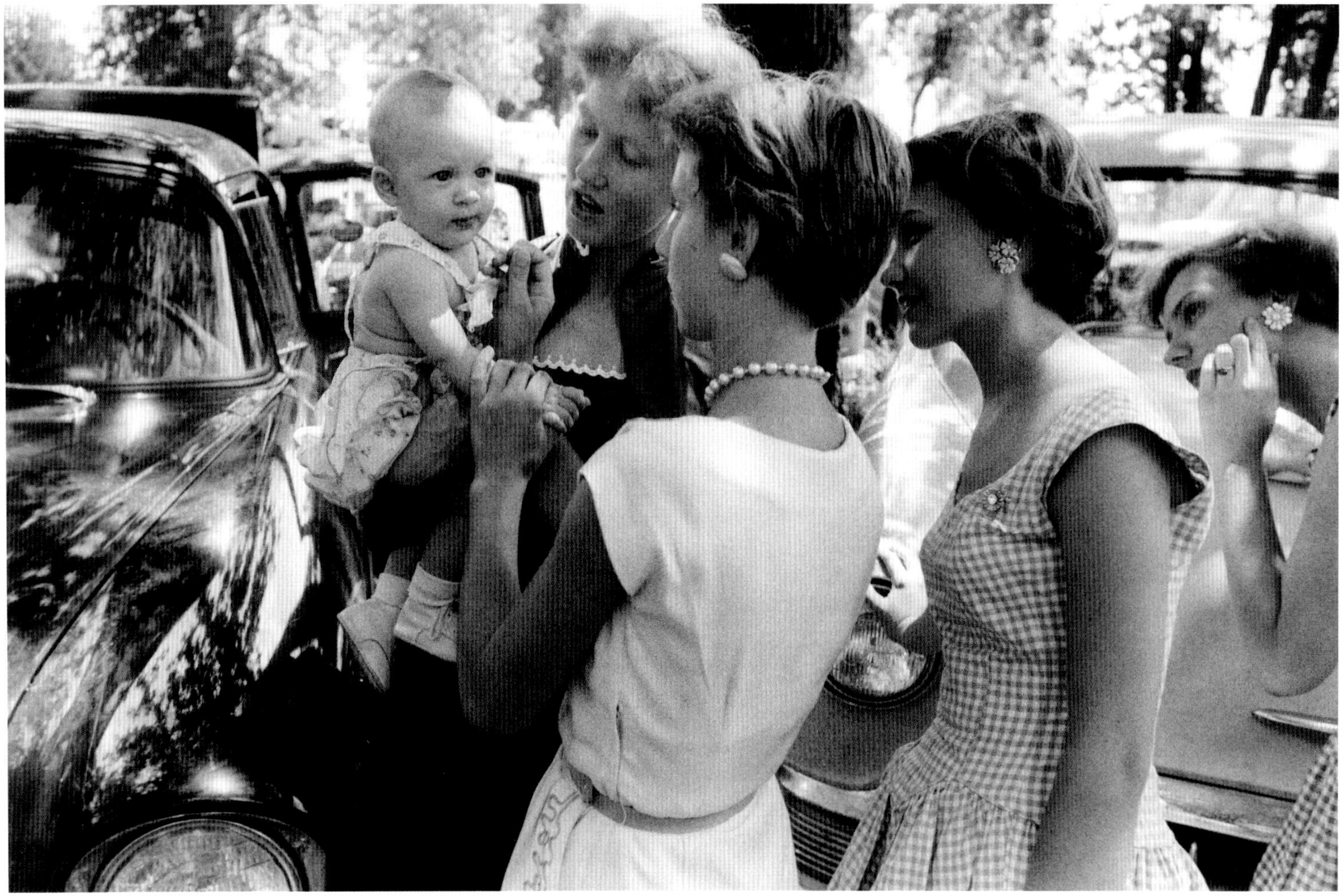

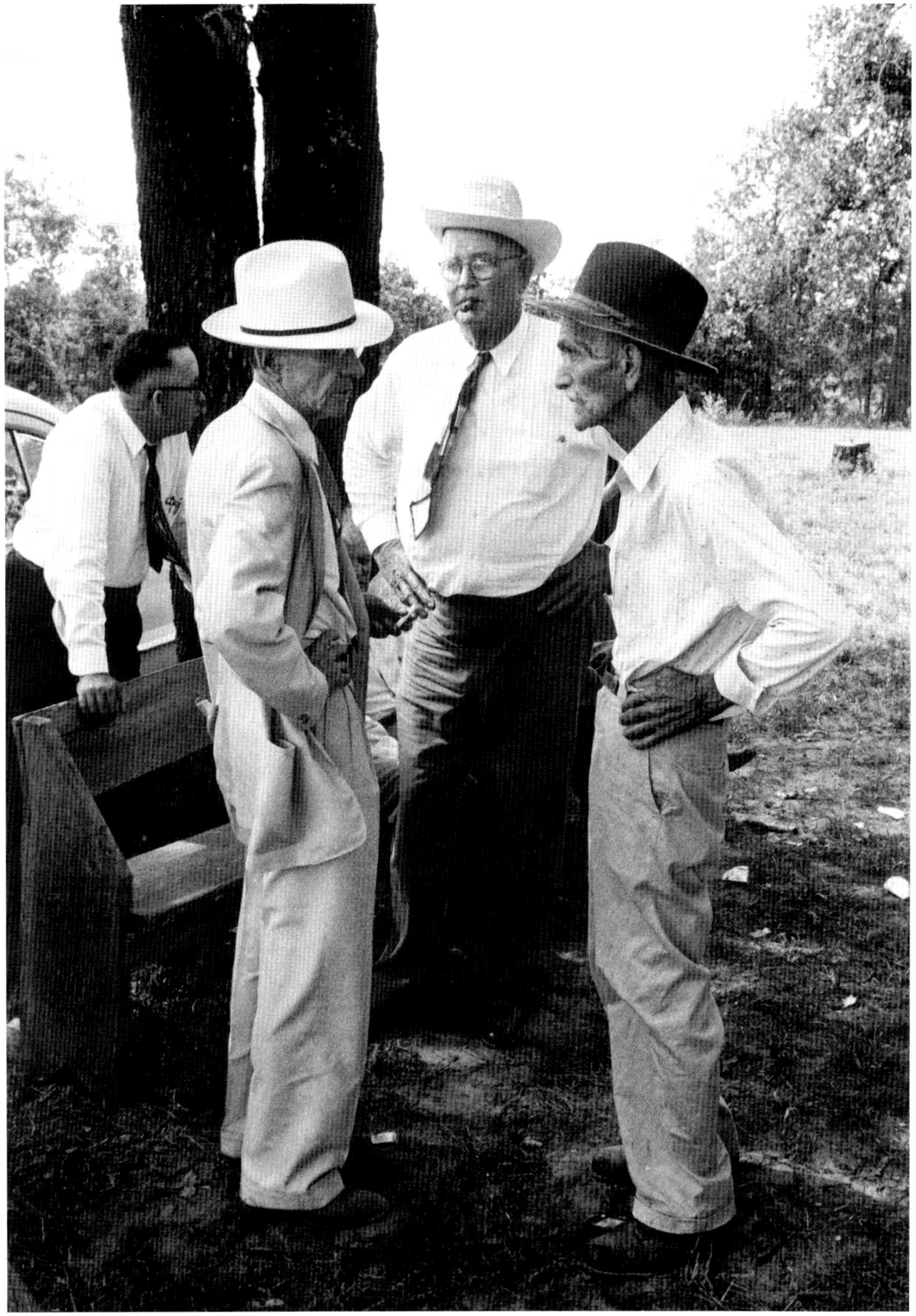

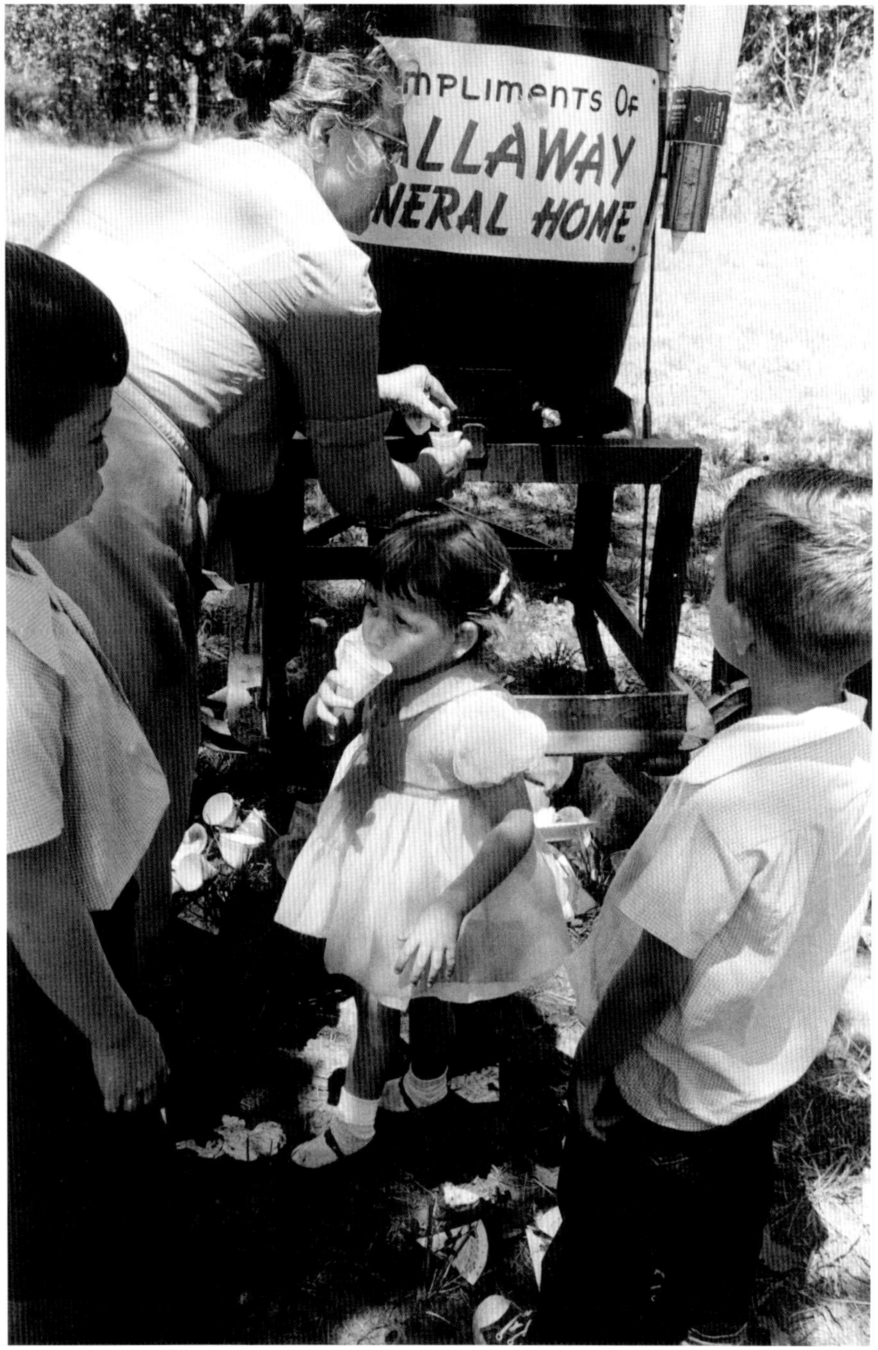
MPLIMENTS OF
LLAWAY
NERAL HOME

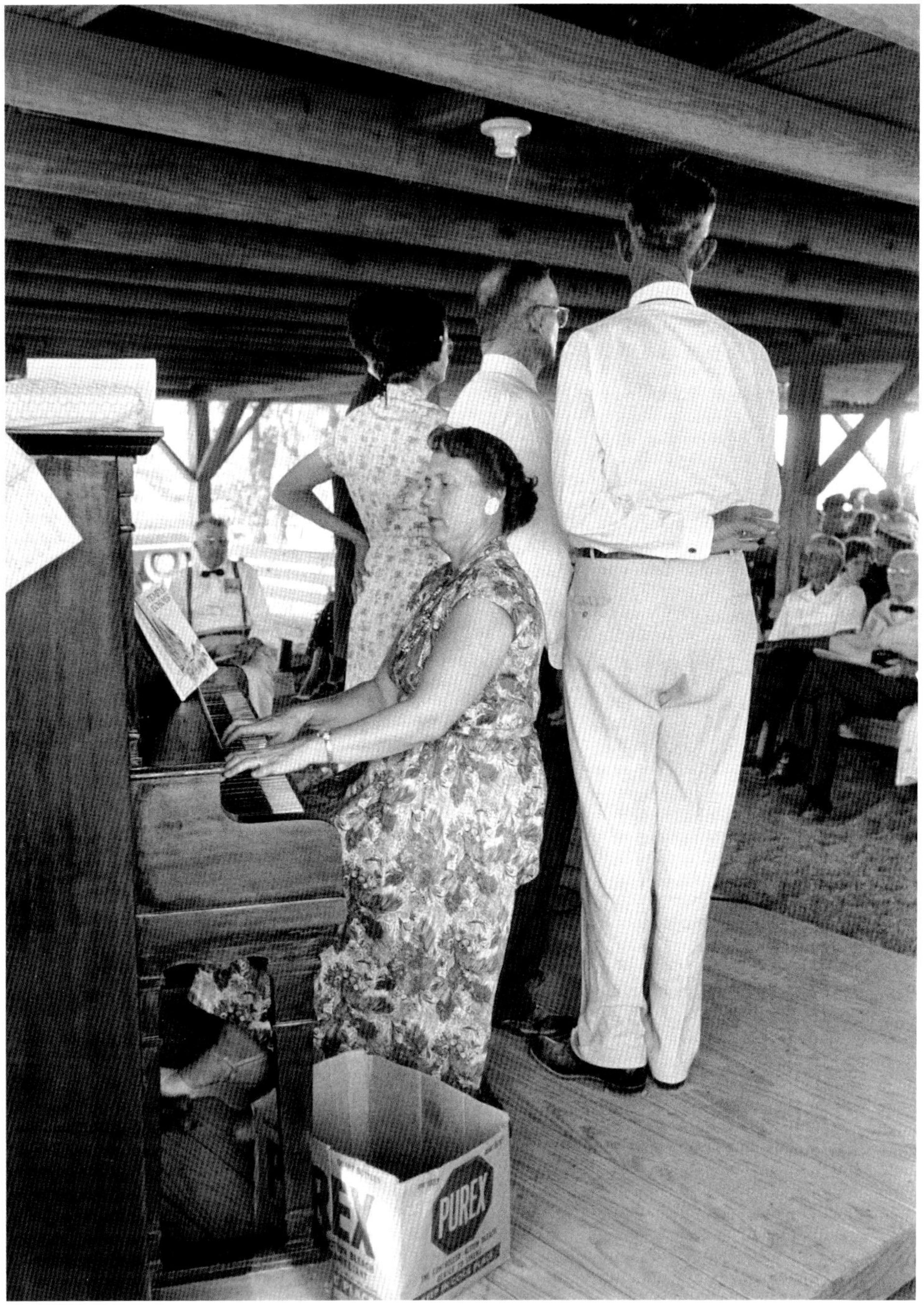
PUREX

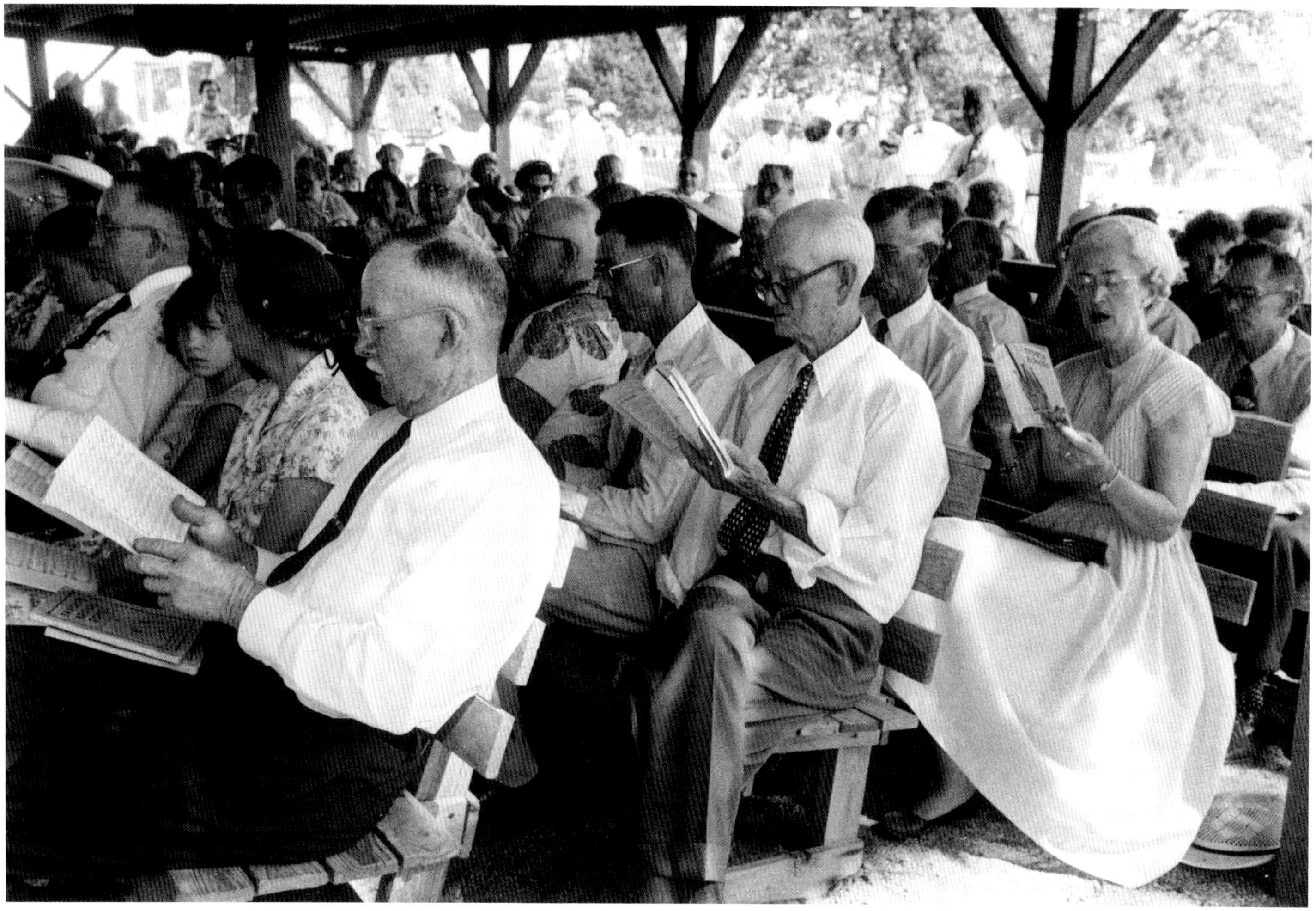

Bible School

Art shows, it doesn't tell. It's not didactic. The documentary photographer is the quintessential artist. The whole of photography itself may be laced with the subjective, with the propensity to tell, to be didactic. For the photographer chooses the subjects and often seeks to instruct them, frames and composes the image and often seeks to arrange its circumstance, then crops and alters the outcome. Documentary photographers face a fundamental difference: they control only their camera, never the world they are photographing.

Art is a process, not a product. The documentary photographer is the quintessential artist. Living in the moment, camera always at ready, trusting on life experience more than on technology, always trusting intuition, trusting the gut feeling. Trust the process, and the product will take care of itself.

Art is the message, not the artist's messenger. A woman uses the church bulletin to fan the baby she's holding in her lap. The pianist at work, the boys and women sing, the men listen. Under the upraised arms of the Savior, the Bible school leader begins the Pledge of Allegiance while the boy holding the church flag says something to the pretty girl who's not thinking of flag or pledge or Bible. Then she laughs with her girlfriends, maybe about the boy. Then the girls, young and beautiful and happy, swirl their skirts in the churchyard. Photographs saying more of value about religion than is ever delivered from the pulpit.

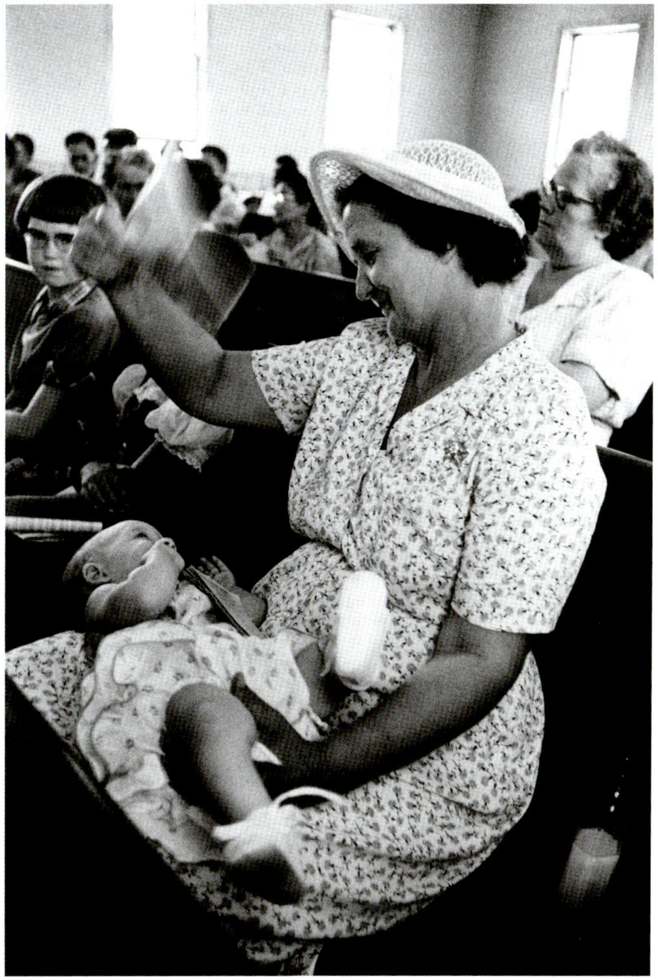

Guy Gillette has never considered himself an artist, has ever only intended to make good pictures. Which is what makes him such a brilliant artist.

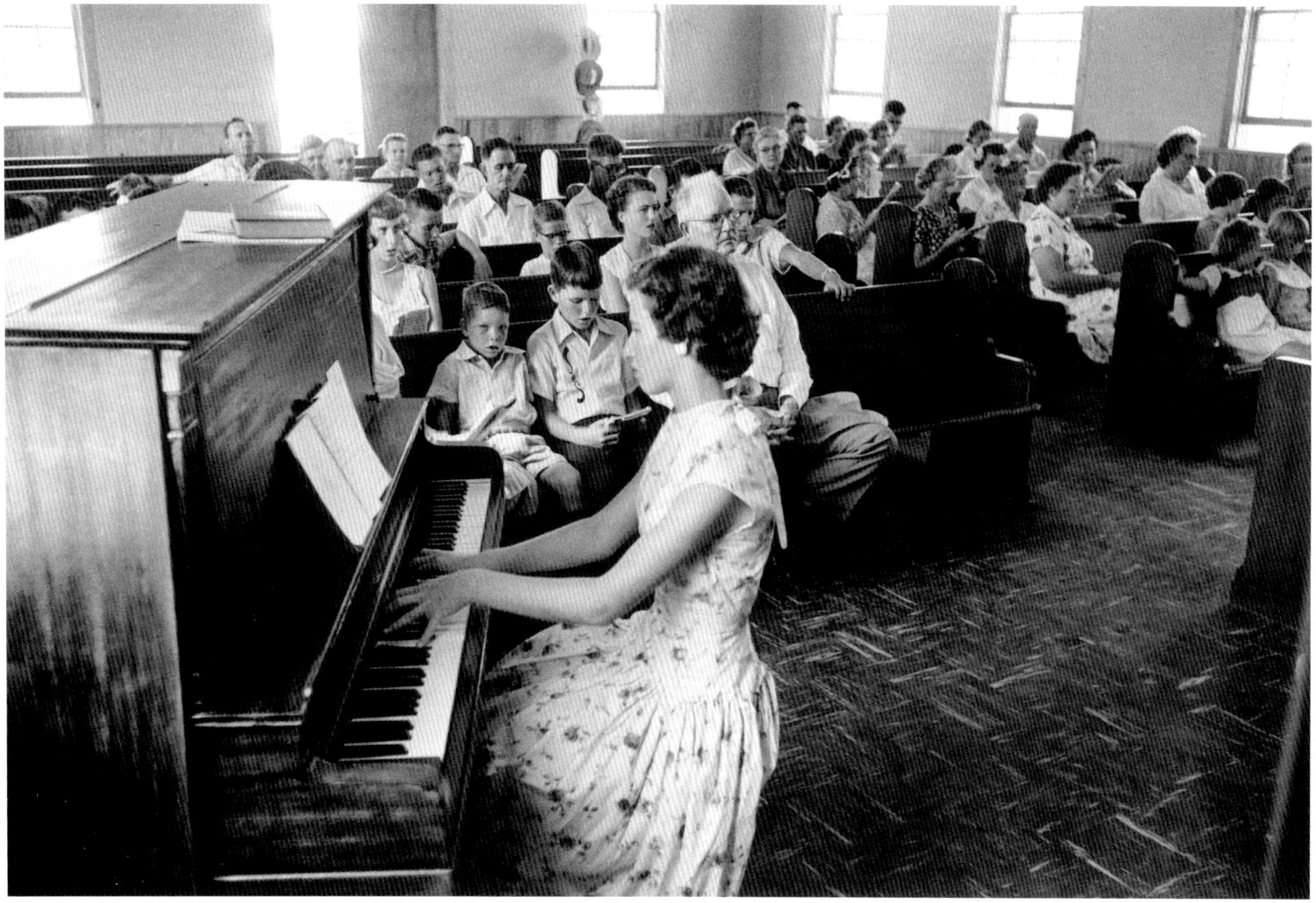

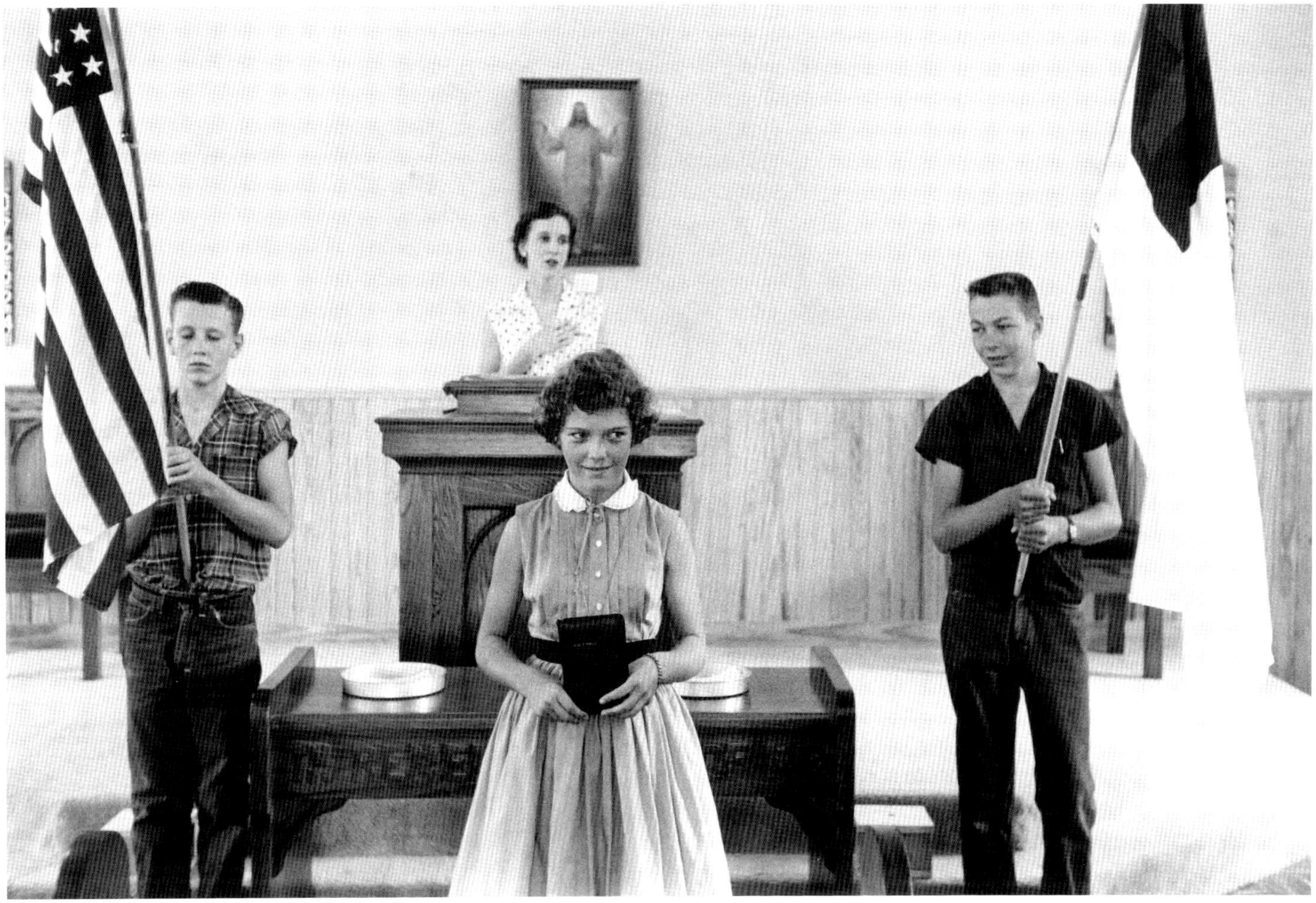

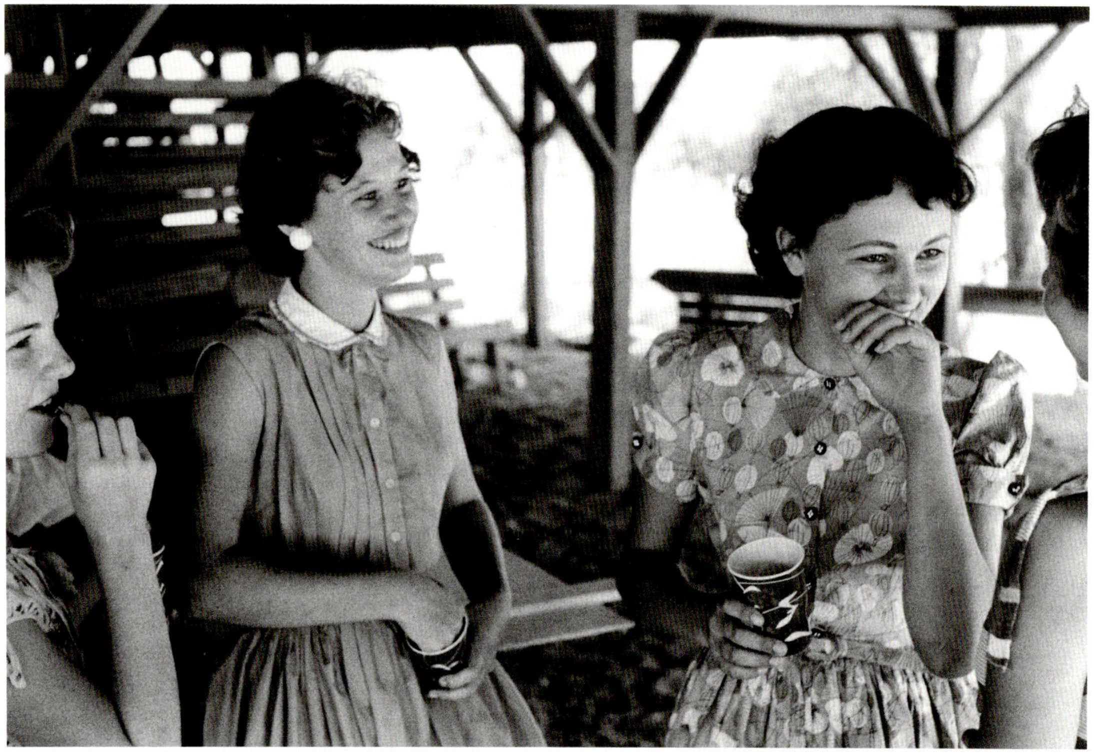

Though photography is often called art, it has always seemed to me that I have wanted to be artless: to practice a craft free from cunning, from artiness; a craft that is not artificial but, rather, simple and uncontrived; to record visual information; to be a documentarian, not an artist.

GUY GILLETTE

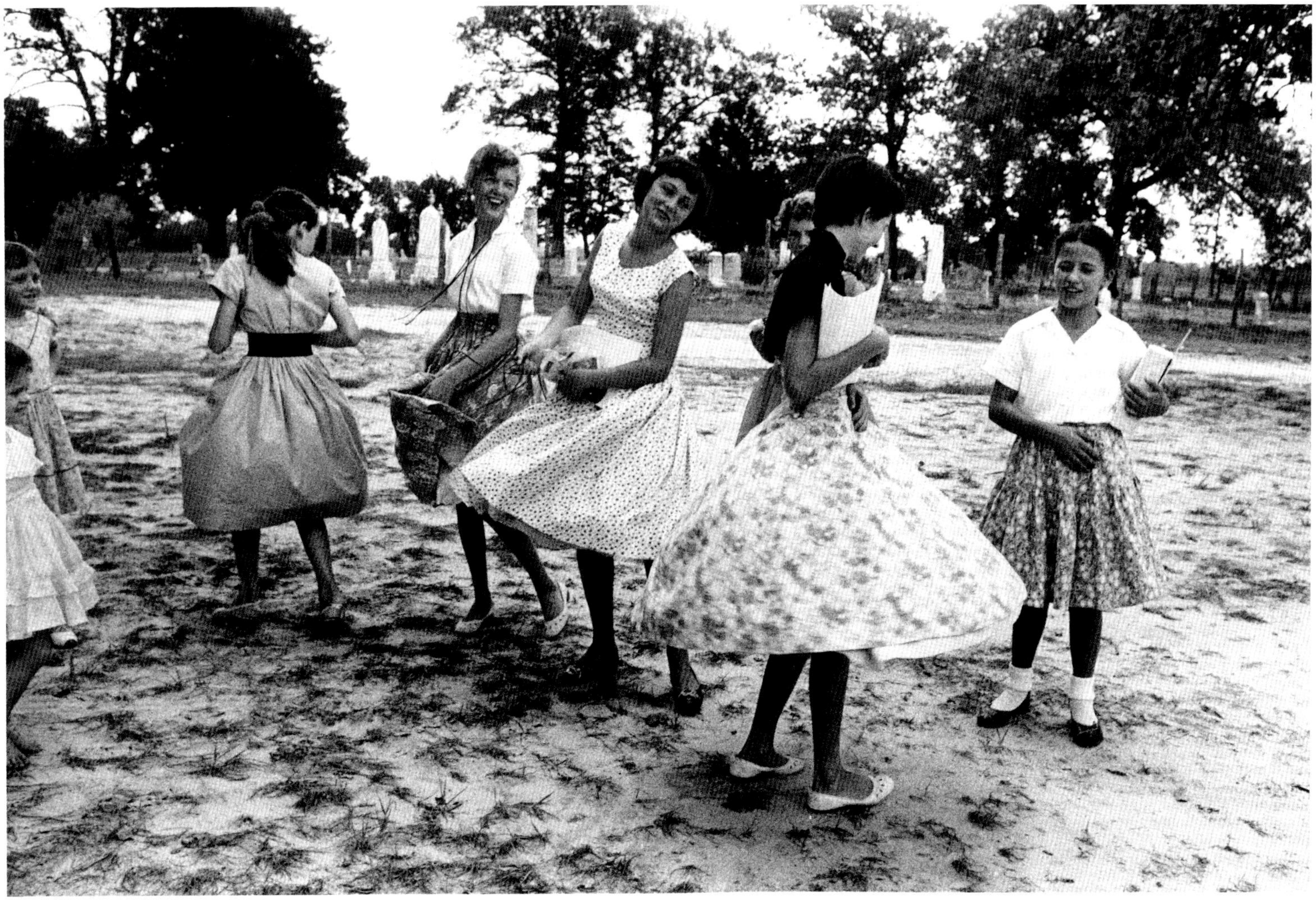

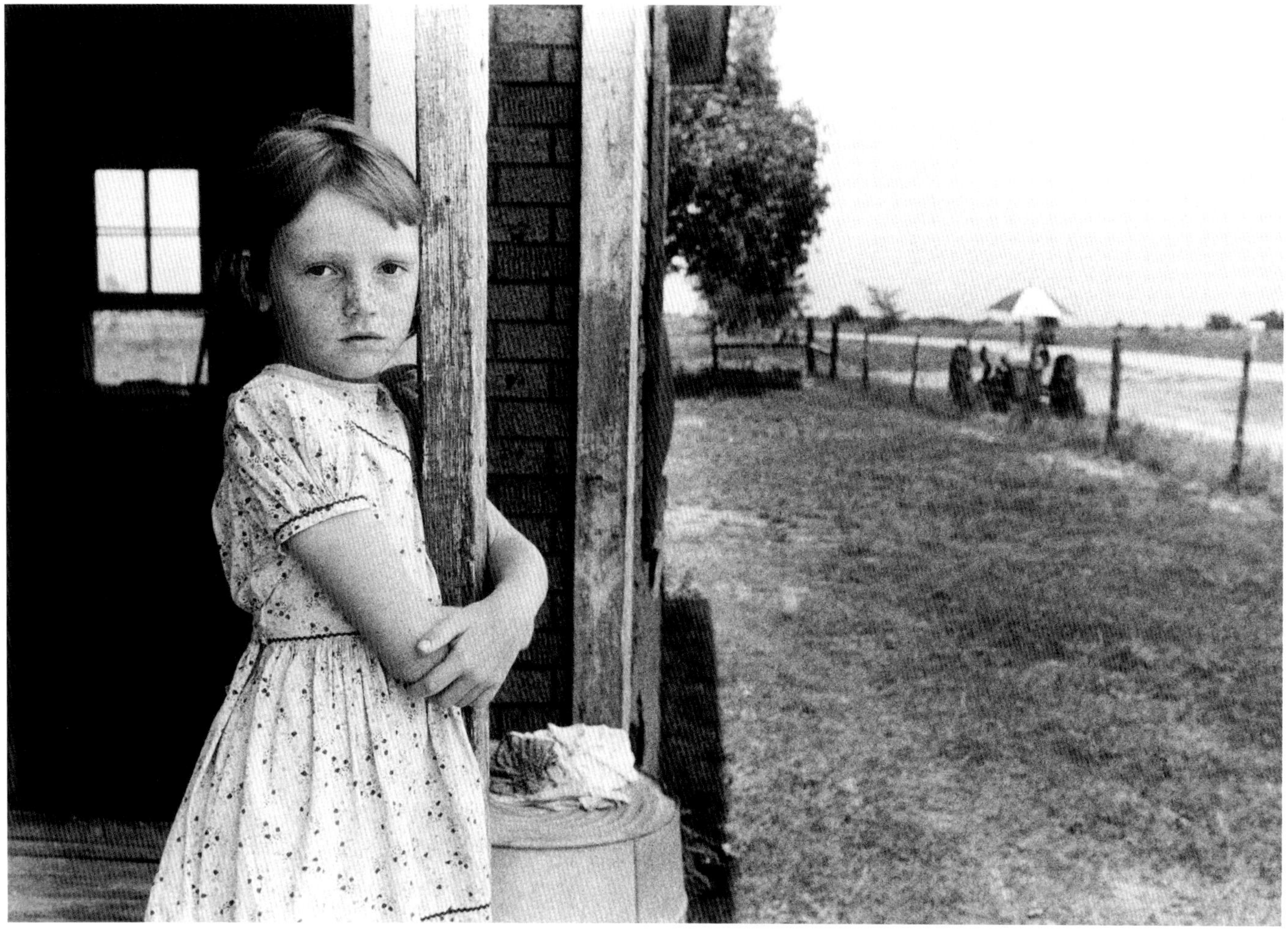

Crossing

This is the crossing over Gail Creek. The channel runs deep, some thirty feet down. After a good rain the water will race through on its way to join the Trinity River nearly fifteen miles south. Pipp Gillette and his brother, Guy Porter, and another Guy, this one the cowhand Guy Pipes, the three of them ride west, headed toward the cattle pens a quarter-mile up the road.

Behind the riders and to their left, hidden by big bull pines and dense brush in between, stands a dance hall. An African American dance hall, on Porter Place land. Another hundred yards east around the bend, on the same side of the road and fronting it, stands a two-story house with a commissary downstairs and living quarters up top. Hoyt and Lucy lived there, raised Doris there.

Keep going east, around another bend, a short piece after the road turns south is Center Grove, a community founded after the Civil War by black freedmen. All the land, all the farmsteads surrounding the Porter Place, belong to African Americans — the Runnelses, the Wootens, the Watsons, others. They are customers, employees, neighbors, friends.

And dancers. In the thick, dark night there will be a patch of sky straight above the pasture that surrounds the Porter house, a small arc lit up by the Milky Way. Maybe tonight it's Lightnin' Hopkins singing "Baby, Please Don't Go," moving couples around the floor, the stars crossing overhead, the Porter family drifting off to sleep.

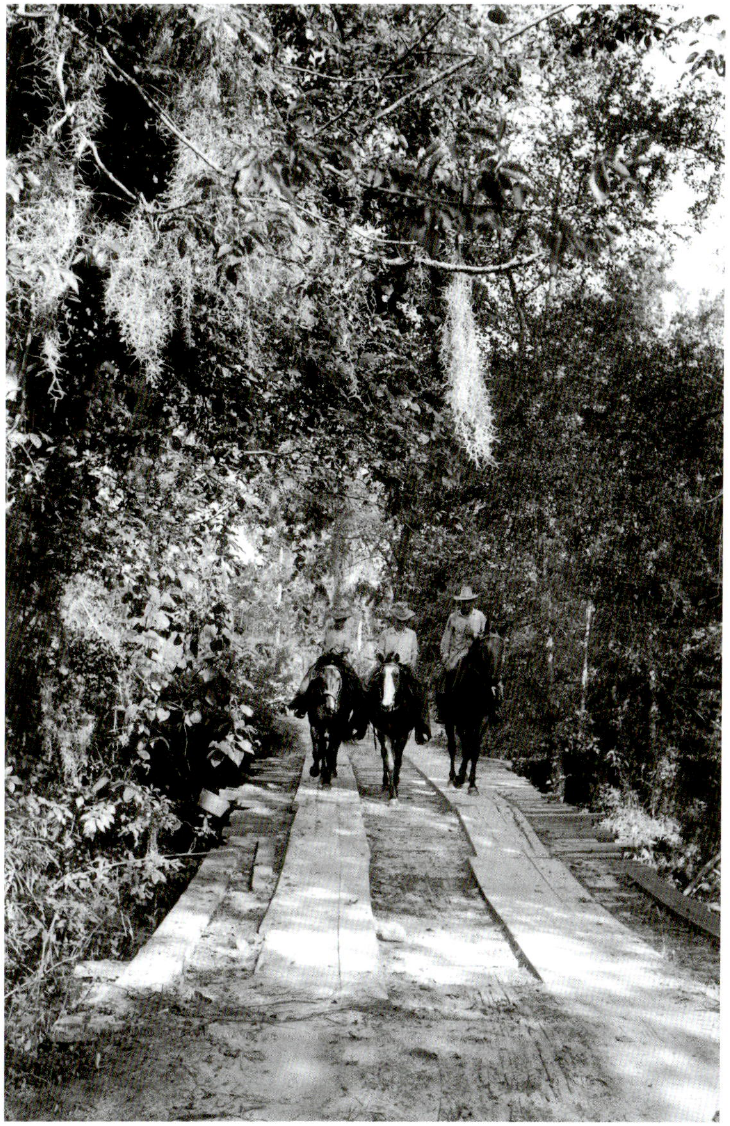

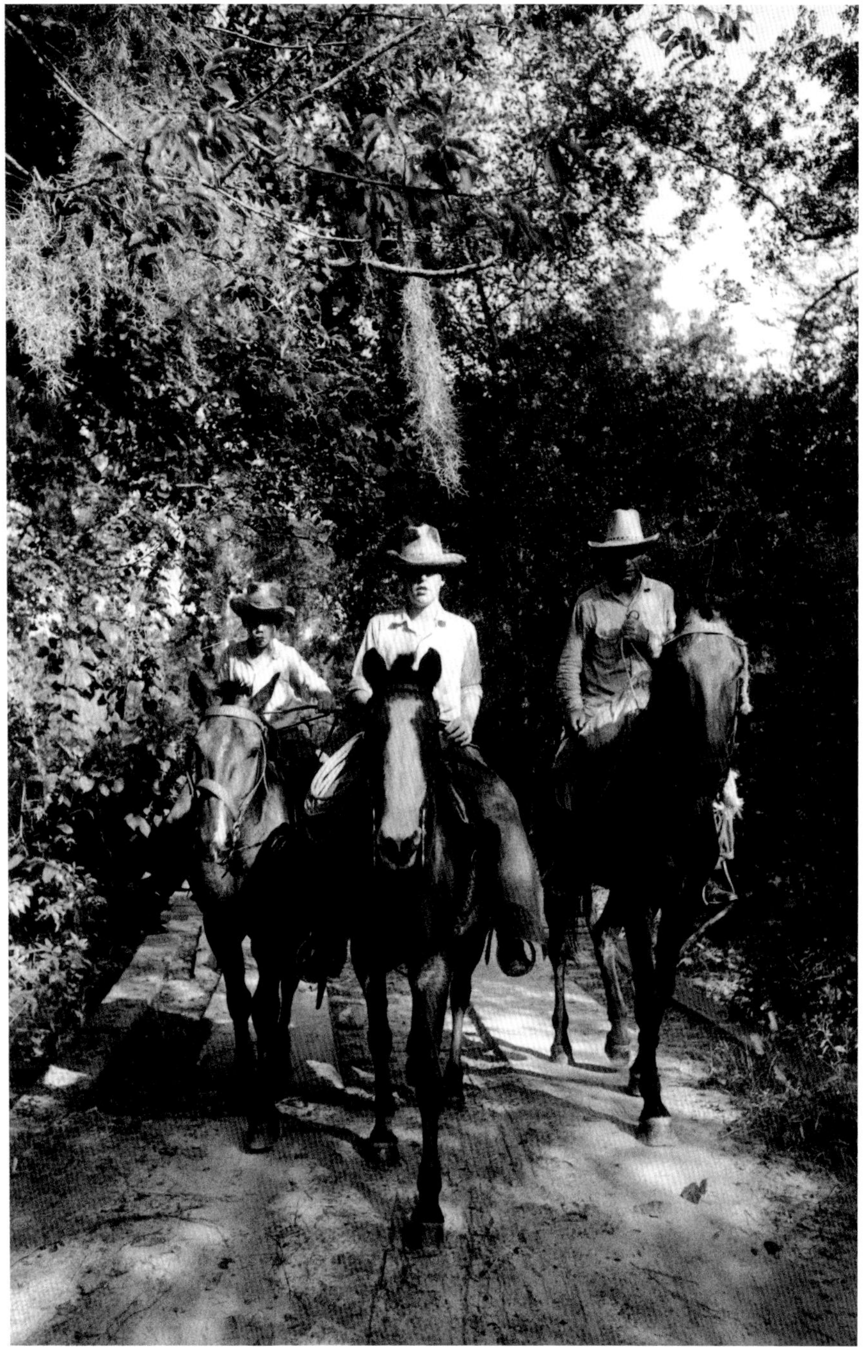

Breakfast on the Range, 1957

There is a problem with this series of photographs. Or not.

By the time they were taken in 1957, Guy Gillette was already an award-winning photojournalist and almost a decade into documenting this ranch and this family. Yet these breakfast scenes are staged. Consider the details. It is well past sunrise, the shadows are short, nearer to noon. A clean skillet lies alone in the grass, almost hidden. No other supplies or cooking gear can be seen — unless a long-handled axe counts as cutlery — yet the cowhands hold ceramic coffee mugs and saucers, dishes more at home in the cafe than on the range. And it is rare that horses are saddled before breakfast, rarer still that they would be allowed so near the camp, especially without hobbles.

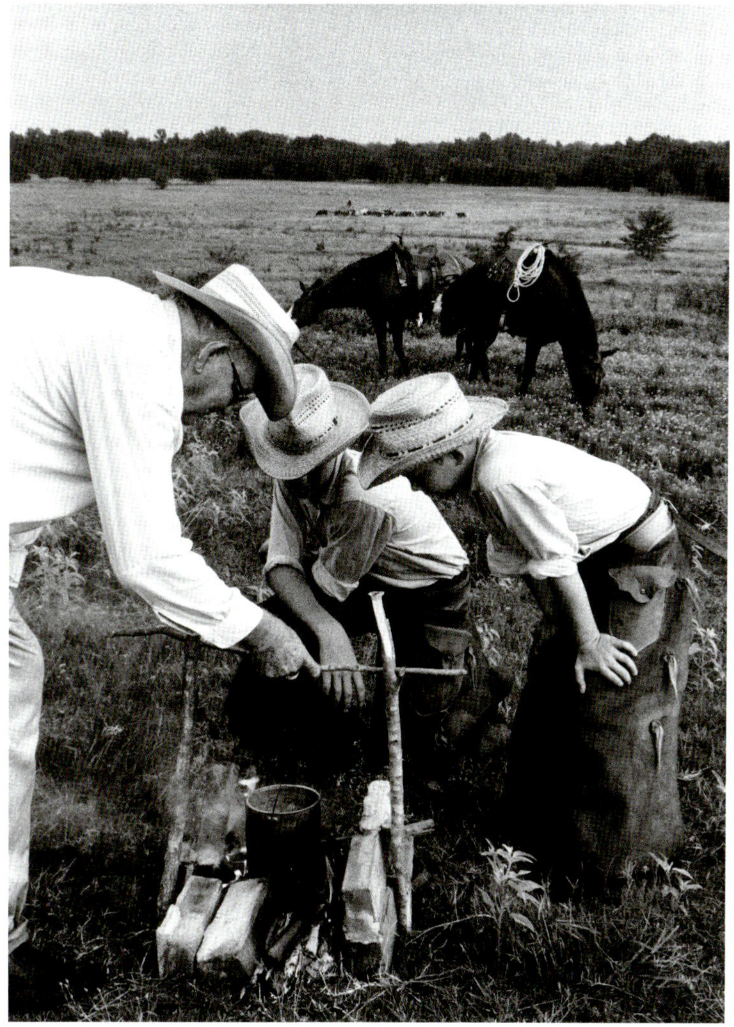

We see intuitively — and have, since photography began — that a photograph is a documentation, the evidence of a specific instant in a specific place amidst a specific action. We believe the old saw, a picture is worth a thousand words. We prefer fixity, uncomfortable that digital technology nowadays leaves the image eternally malleable.

Our intuition, however, is tempered with ambivalence and practicality. We don't expect that all portraits be candid snapshots, or all school graduations, or all birthday-cake cuttings. We are

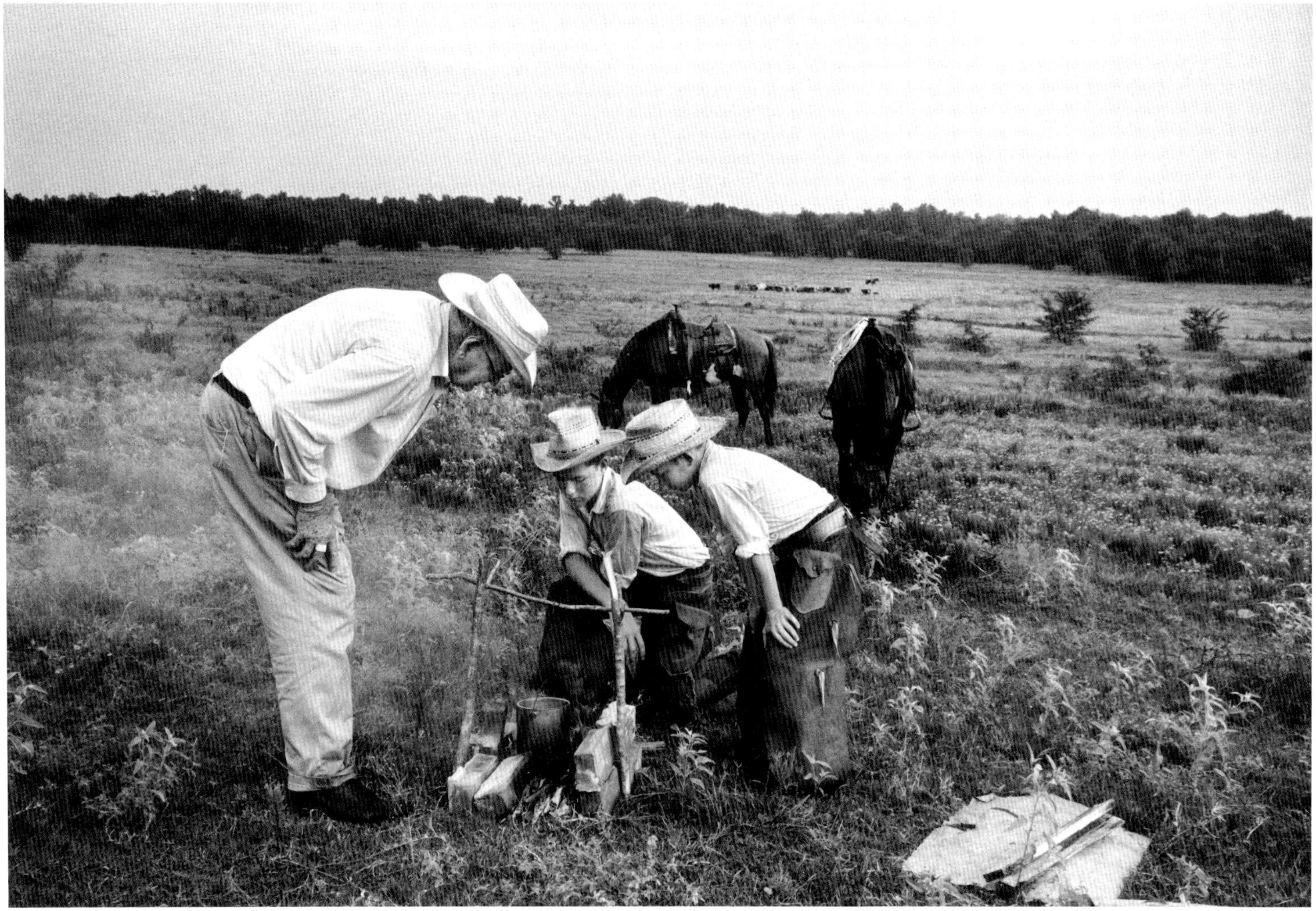

amateurs, but all we need do is pick up a camera and take aim. We have no compunction against setting the scene. Smile. Don't squint. Look this way. One more shot.

We see that there is a story to tell, that every choice the photographer makes is part of the narrative: not just subject and setting but angles and perspectives and lighting and film and shutter speeds.

And breakfast isn't the story here. Consider the details. One generation passing on a way of life to another. The attentiveness and the joy of each. Look at the faces. The photographer tells a story.

Any photograph that tells a true story is a true photograph.

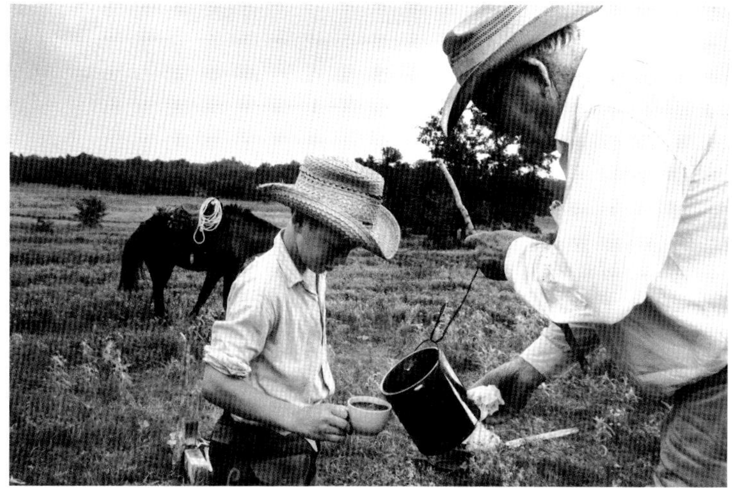

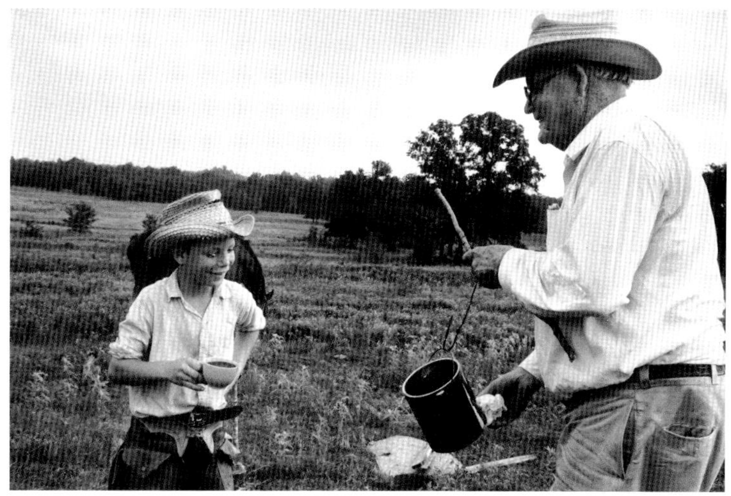

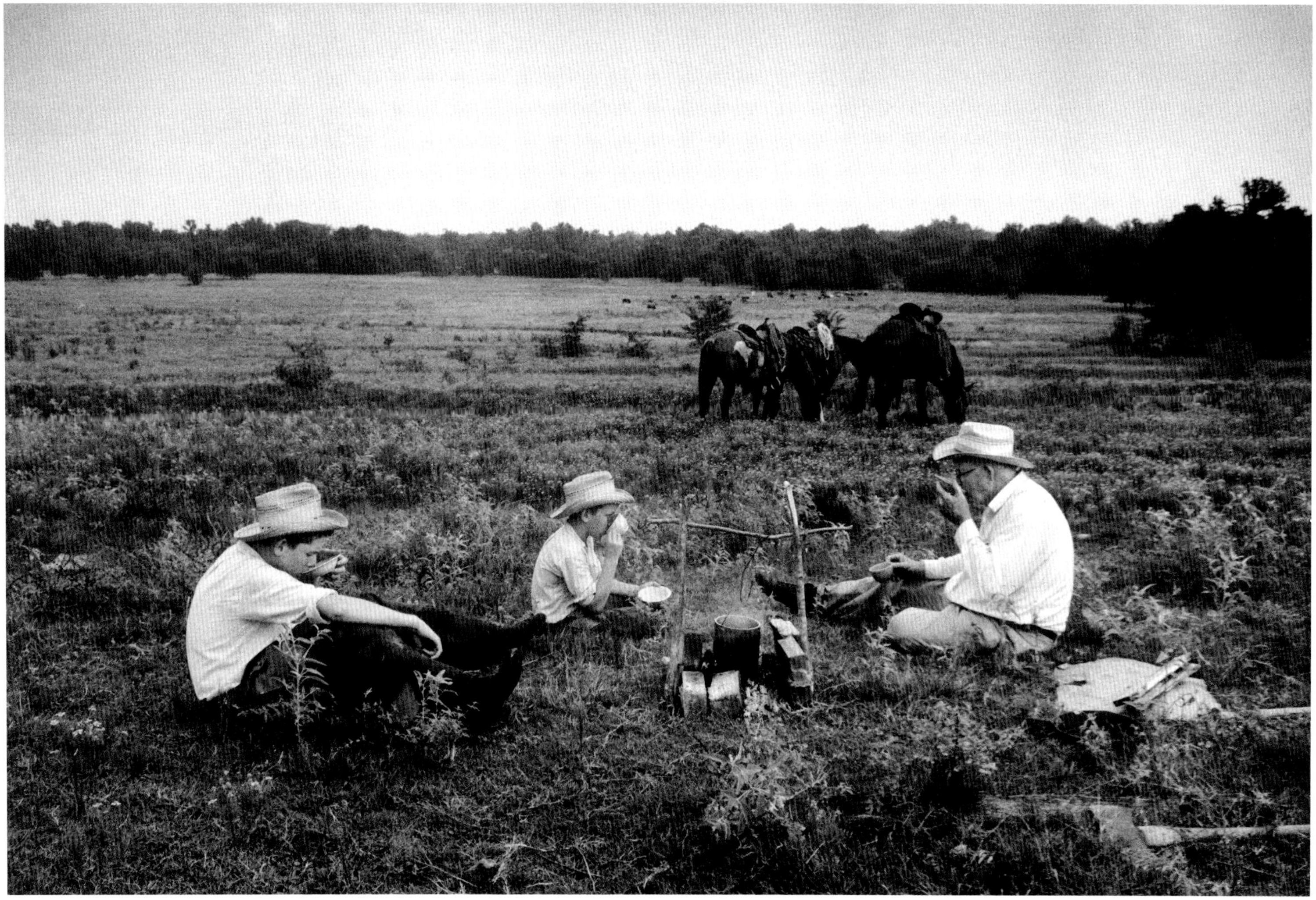

Working Cattle, 1957

You can't push cattle in this country. They might start running, and that's about the worst thing that can happen here. If they make it to the woods and the brush before you do, they'll sull-up and you'll play hell getting them out. Once they get in the thickets, it's damn near impossible for a cowboy to work them horseback.

Farther west, farther north, when you leave this edge, this boundary between the American South and the American West, the land dries out and opens up. It grows expansive. Instead of the forty inches of rain that can be expected year in and year out on the Porter Place, a rancher in those further regions is more likely to get only a third of that. Out there, it's mesquite and shin oak, prickly pear and yucca and sage. Out there, it's all bunch grass, higher in protein but less of it per acre and with a short, short growing season. It's a long way from water to water out there. In that open country, the carrying capacity — how many head of cattle each section of range can support — drops quickly.

Out there, ranges and ranches have to be big. When you move cattle out there, there's distance to cover. You have to push 'em. But back here at the Porter Place, in the square pastures that began as antebellum cotton fields carved out of the woods, a cowboy has to be constantly attentive, has to use his southern dogs that can bay-up the cattle, keep them bunched up, keep them contained.

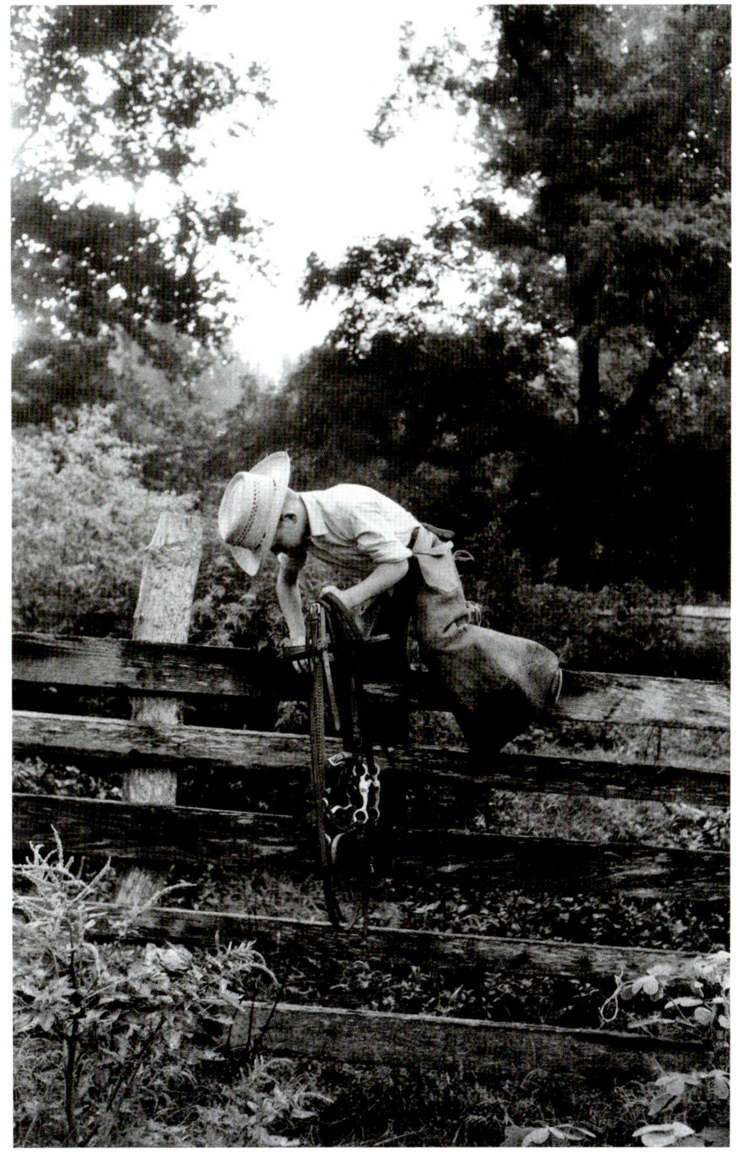

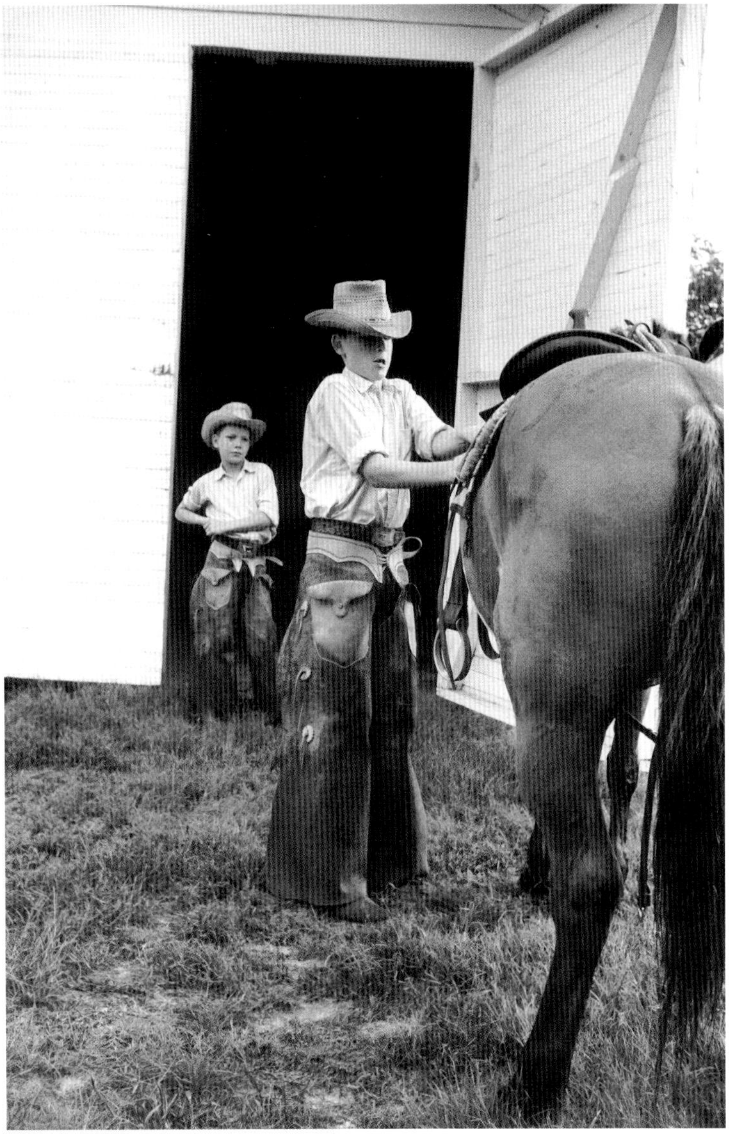

The cowboy has to be big out there. And the cowboy has to be big back here, too, everywhere there are cow works to be done, regardless of the size of the spread. Not in size, not in age. Big where it counts, in heart and mind and experience and have-to. Hoyt, Dayton, Claude, they've schooled these two boys well. Guy Porter and Pipp sit taller in the saddle now, in more ways than one. They know how to be in the right place at the right time. They know better than to push the cattle.

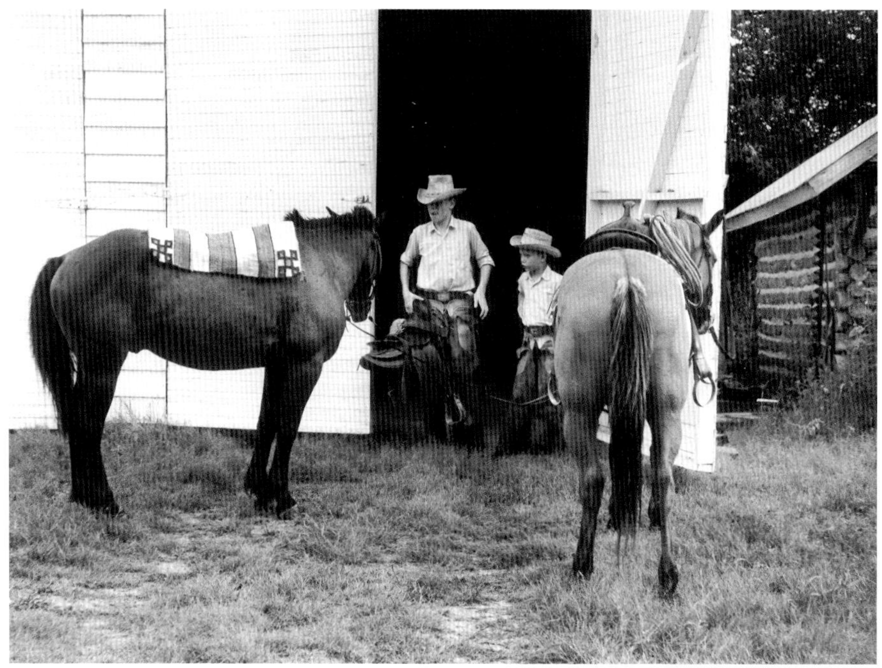

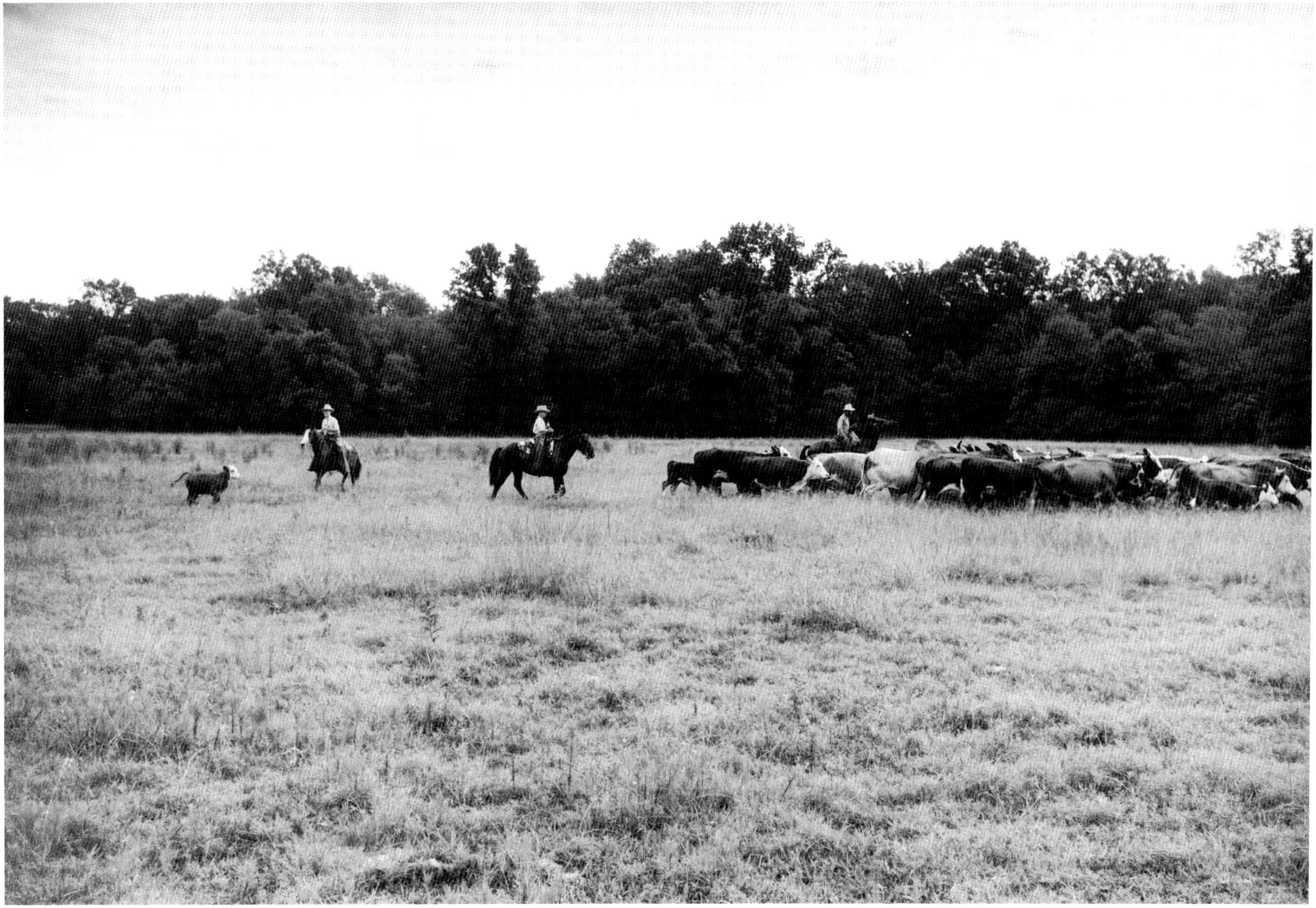

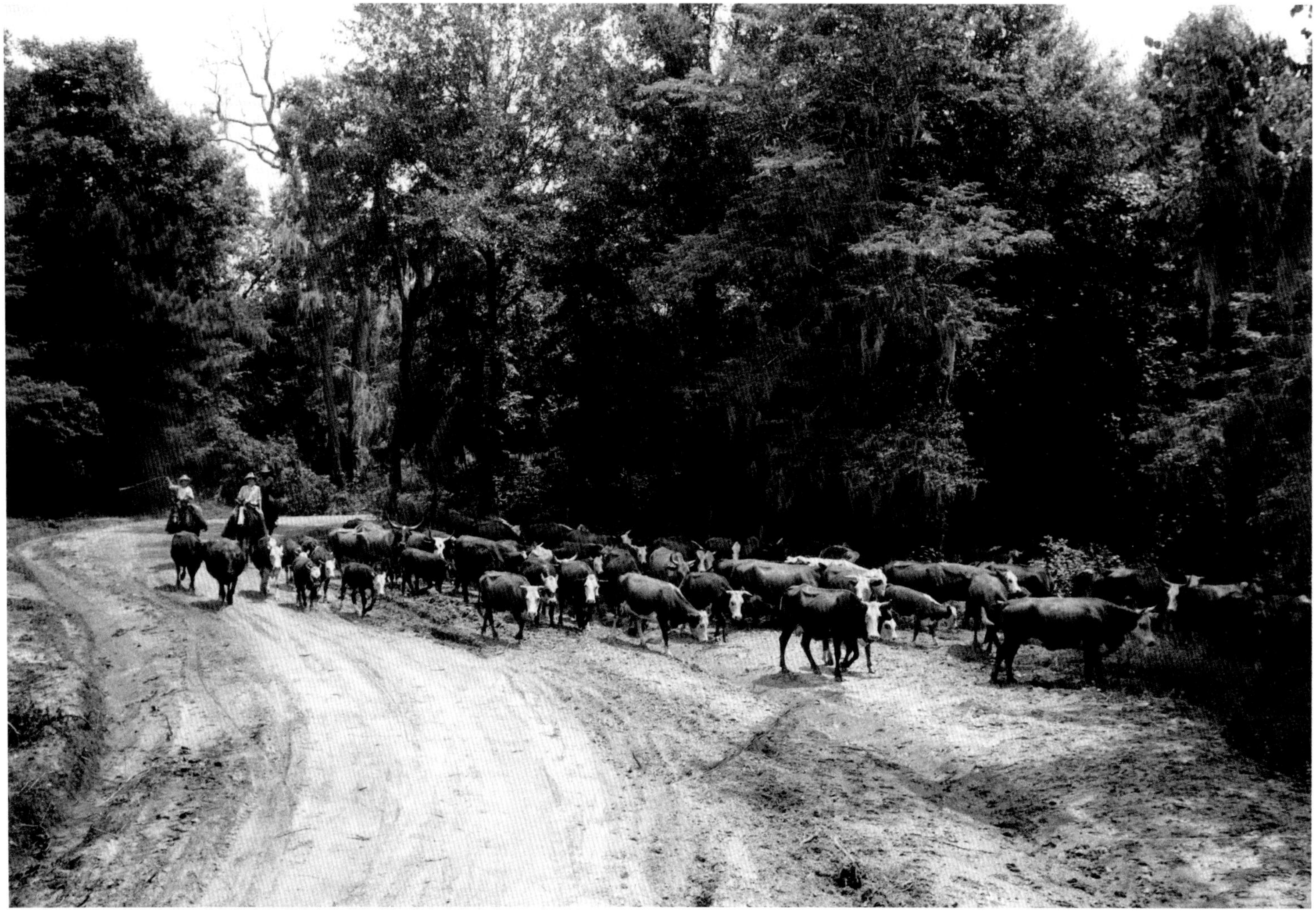

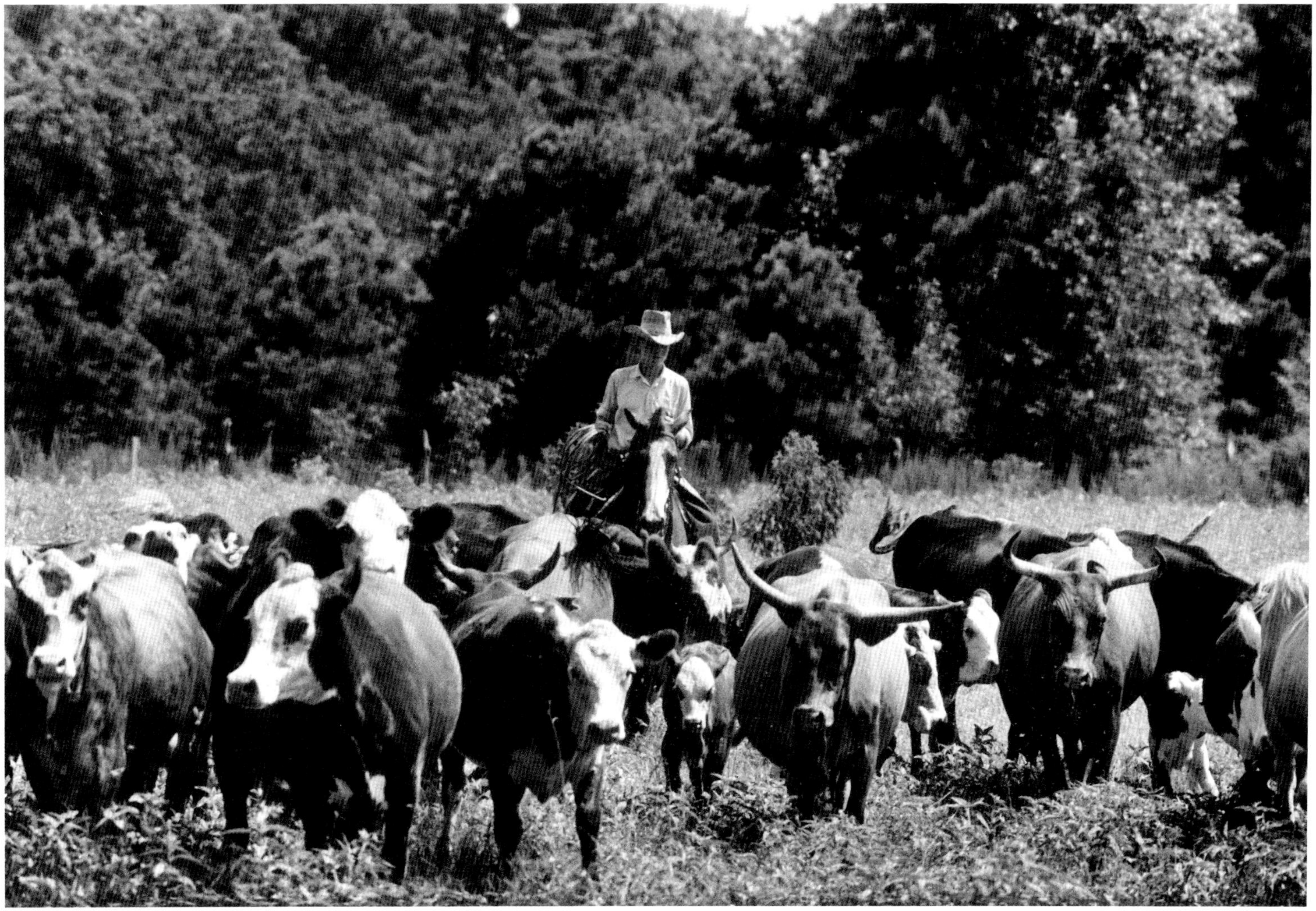

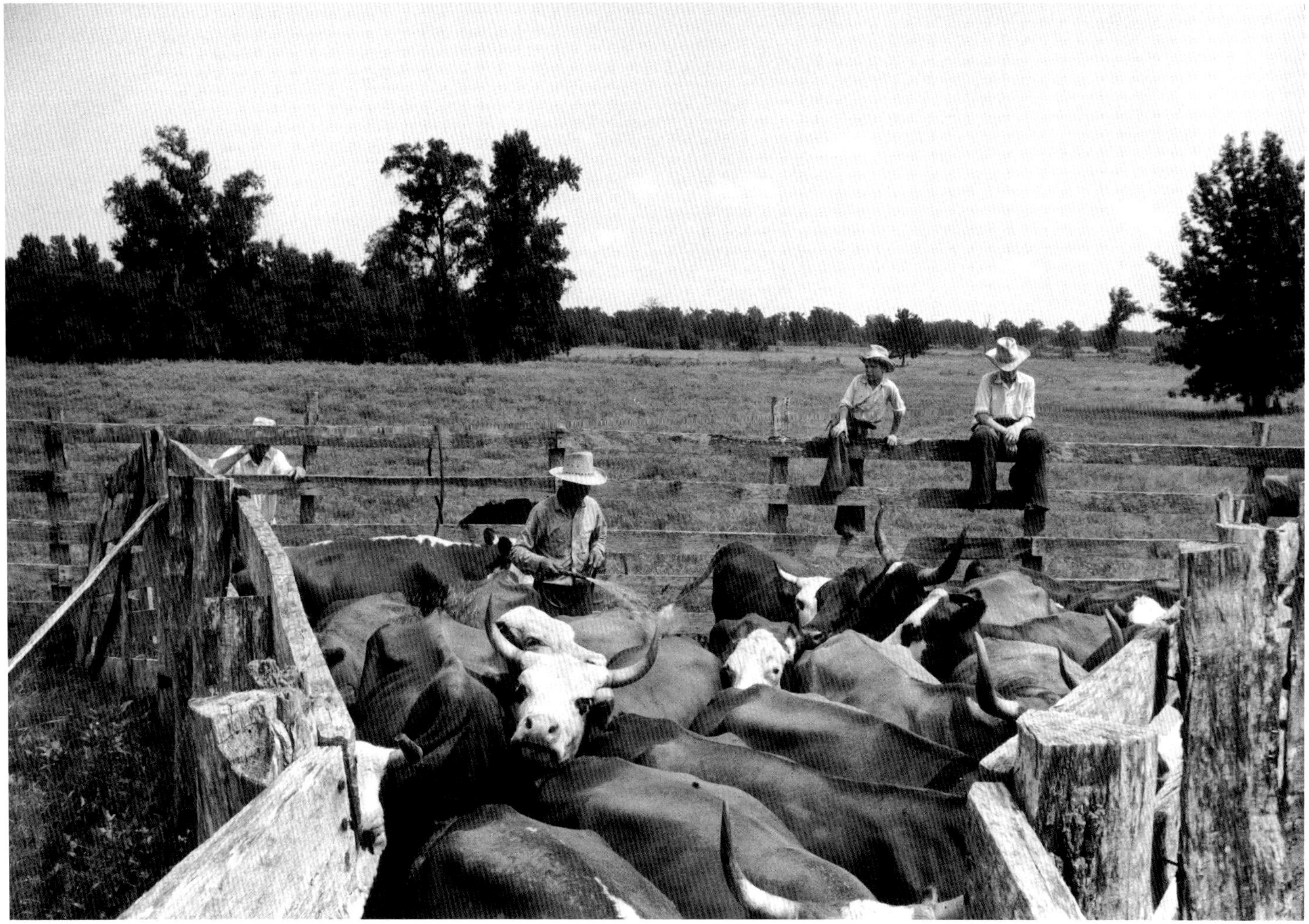

Water Hole

Just west of the bridge over Gail Creek there's a big pasture that opens off the road and runs gently southward. Near its northern boundary, there's an artesian water hole. Grass and water, water and grass — the cattleman's *condiciones sine quibus non* — without these, there would be nothing. Townies might balk at drinking from a water hole in a grass pasture. But cowboys, even city-bred ones, learn early the need for water, and the ways to get it.

Back in Yonkers, the boys were growing up in the ways of the city. There's different water there. Hoyt listened to "Old Dan Tucker" back in Texas, but the music that played in the Gillette household ranged way beyond, to opera, folk, country, and blues. One early February Sunday in 1964, a whole other music bubbled up out of the Gillettes' television set, a clean, fresh pool in the lives and living rooms of more than a third of all Americans. "She Loves You" and "I Saw Her Standing There" and the chart-topper "I Want to Hold Your Hand": The Beatles on Ed Sullivan. "We had a band from then on," says Pipp.

They were soon known as Big River with Guy Porter on bass and Pipp on drums. The brothers drank full of this new water. It became their livelihood, their passion. They performed all up and down the East Coast, then into Texas. They still came down for ranch works now and again, but the days horseback were fewer and further between.

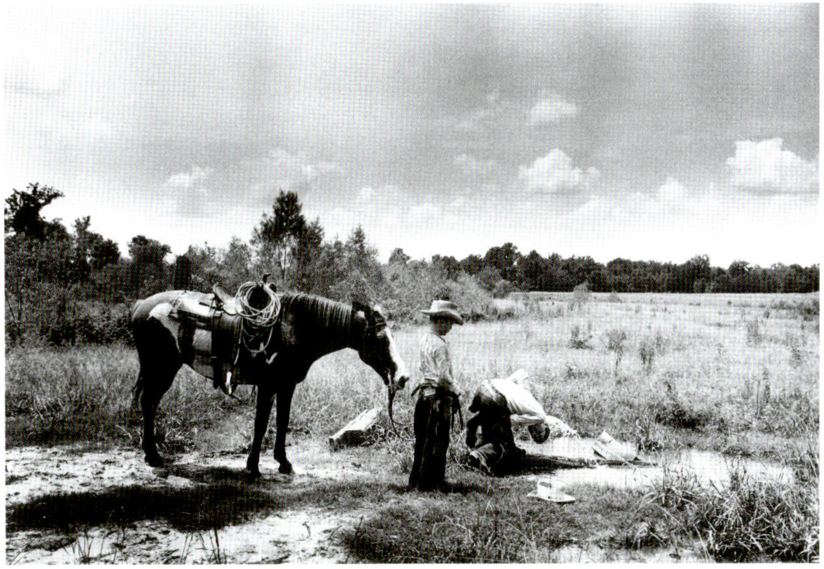

And as all of America was changing in those years, so was the Porter Place. Hoyt leased the ranch to a nearby outfit in the late 1960s. The two-story Porter house burned down in 1968. Hoyt died in 1974, Lucy two years later. For the next decade, there was nothing left to bring back the Gillette Brothers — the name under which they now performed — nothing left to bring them back to the pastures along Gail Creek.

Except in the purely legal fiction of a deed, except in an entry on a county property-tax roll, this family of the land was, for the first time in three-quarters of a century, separated from the land. Not their grass any longer, not their water.

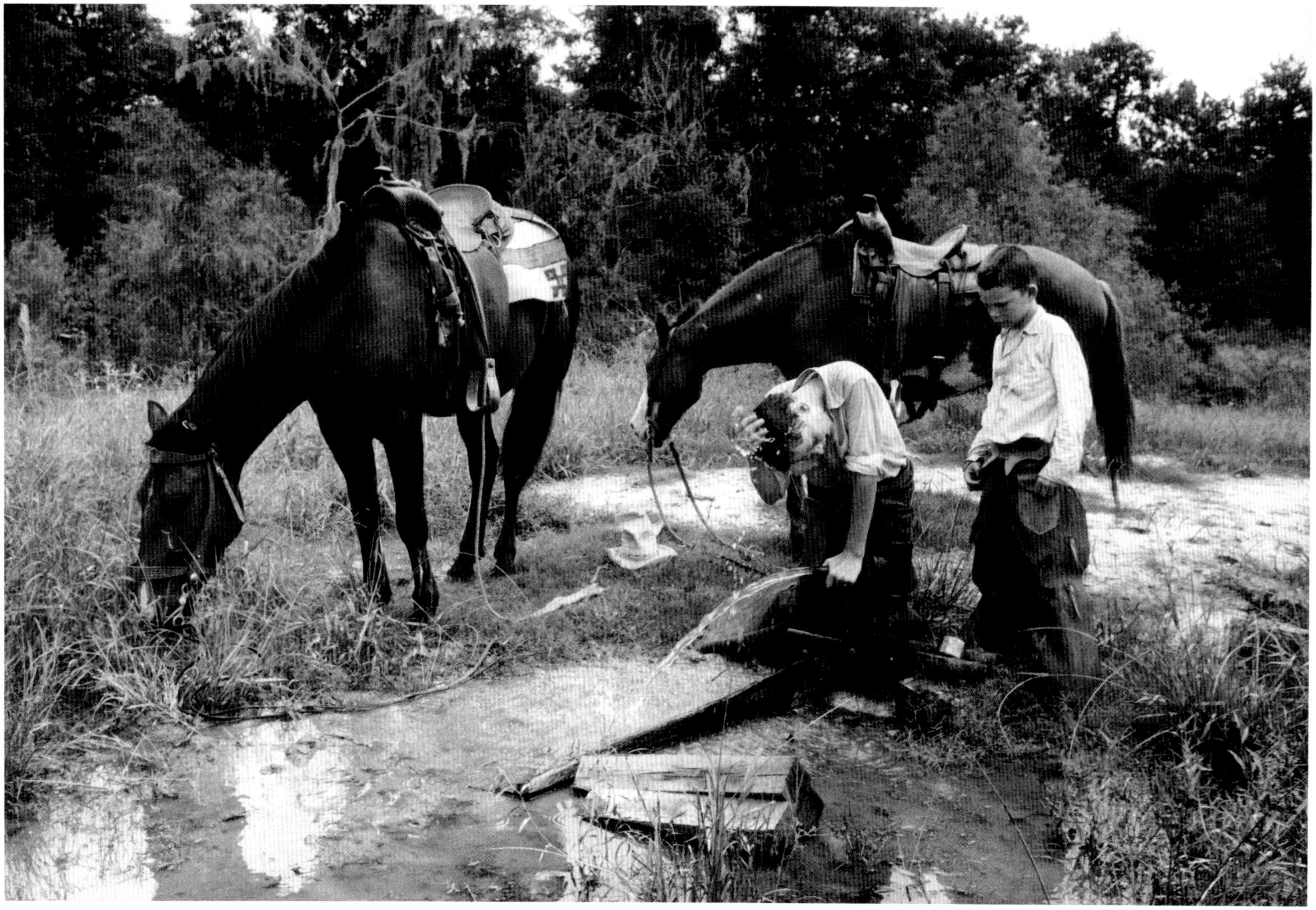

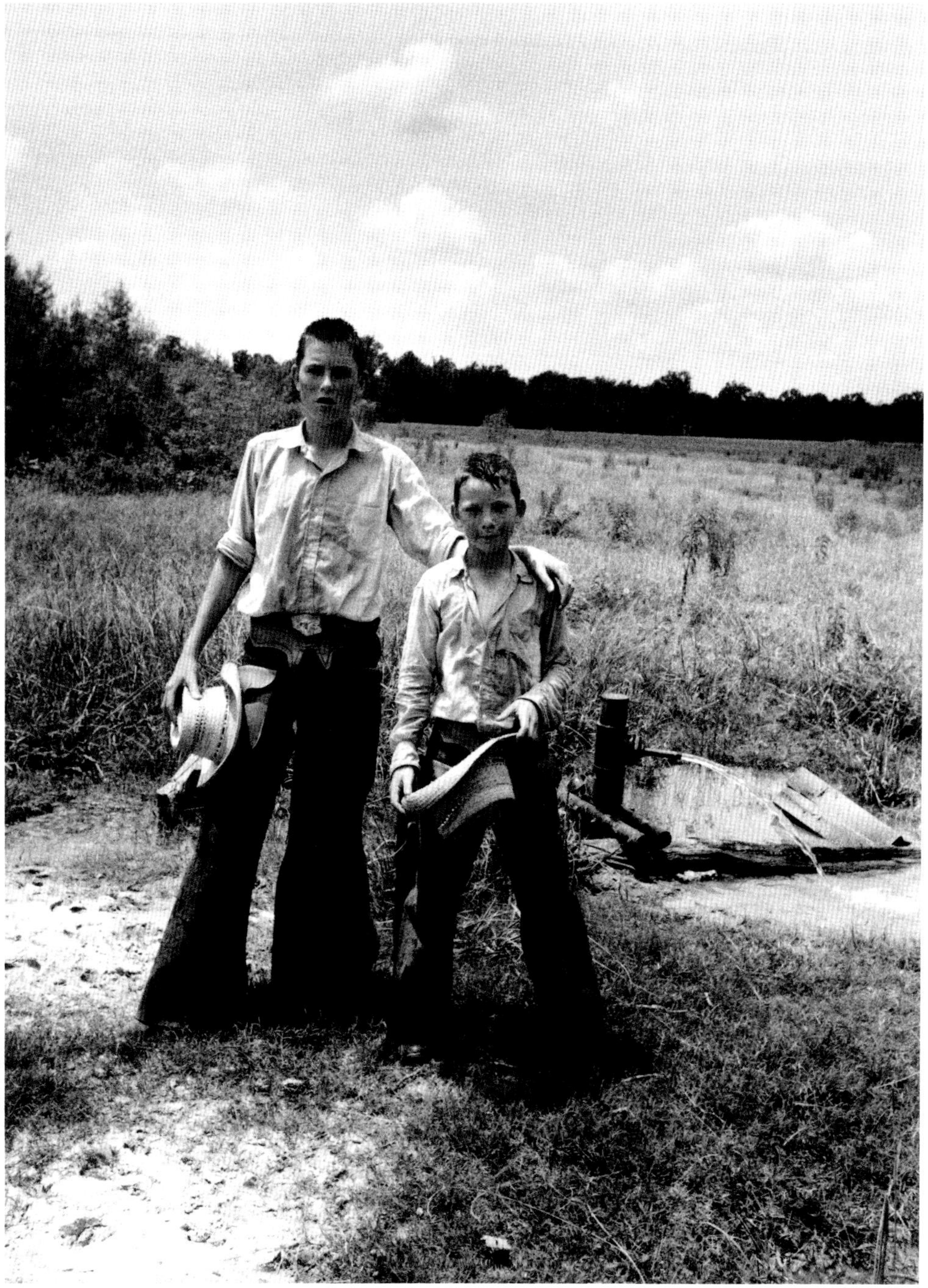

Working Cattle, 1980s–1990s

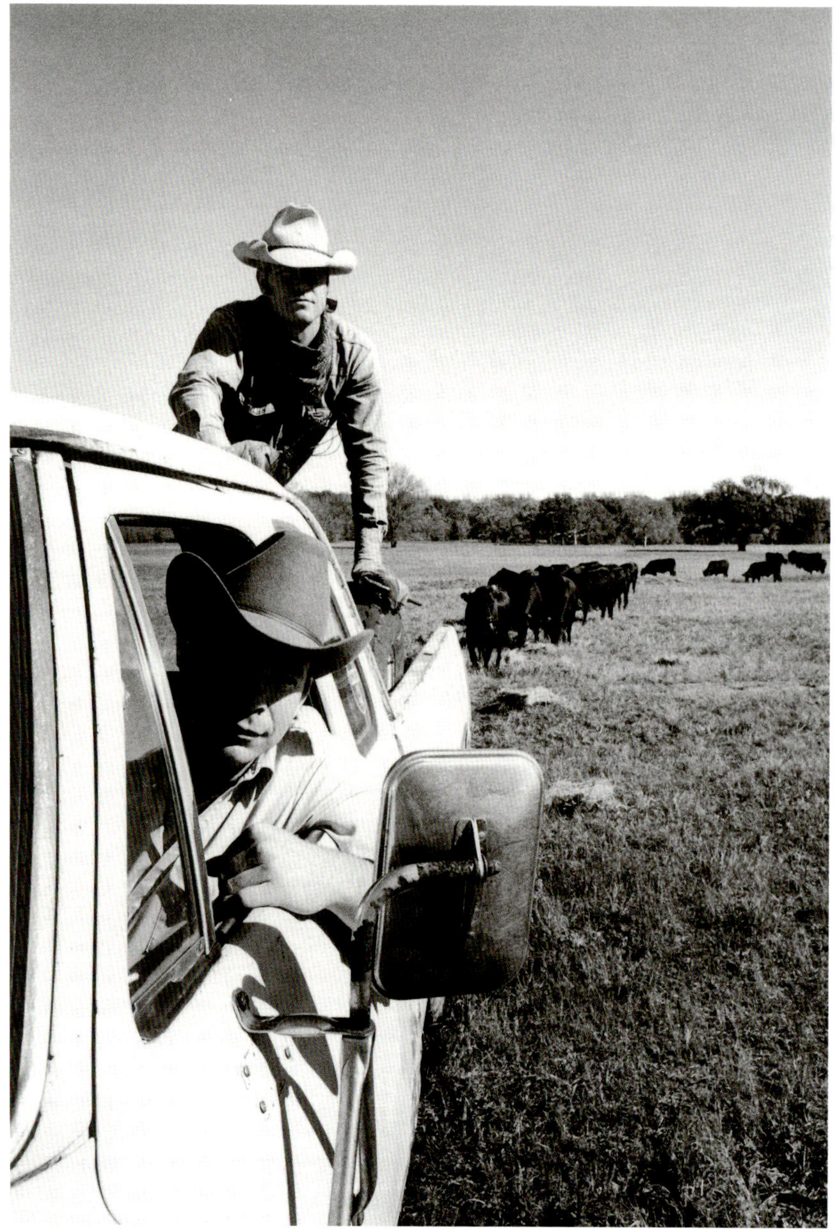

The lease wasn't working. At least not for the family. Cash flow is only one small way to look at land when the land's your life. When the land is only cash flow, there's soon no life in it. The pastures that Hoyt had worked so long and so hard to reclaim from cotton field and pine forest were being taken back by brush and trees. Fences and pens were falling into disrepair. By 1983, when the lease was up for renewal, the land was in such a sorry state that it would have been difficult, maybe impossible, to find a suitable manager. Too much work to be done to get it into shape, too much money to put into it for small chance of small return.

Besides, by 1983 the Gillette Brothers had seen pretty much all the performing world that they were of a mind to see. They'd been on tours as far out as the Soviet Union but increasingly found themselves spending more and more time picking and singing in Texas. The move seemed logical. They returned to the Porter Place, this time for good.

Doris came down for the first few months to help. They all moved into one of the houses that was already there, a small one-bedroom that had once been home to Claude Searcy and his wife and two daughters, then most recently to Guy Pipes, who had been moved, kicking and screaming, into a nursing home not a year or two before the Gillettes came back to stay. The brothers used what lumber and stone remained from their grandparents' house to make repairs. It would become Pipp's home.

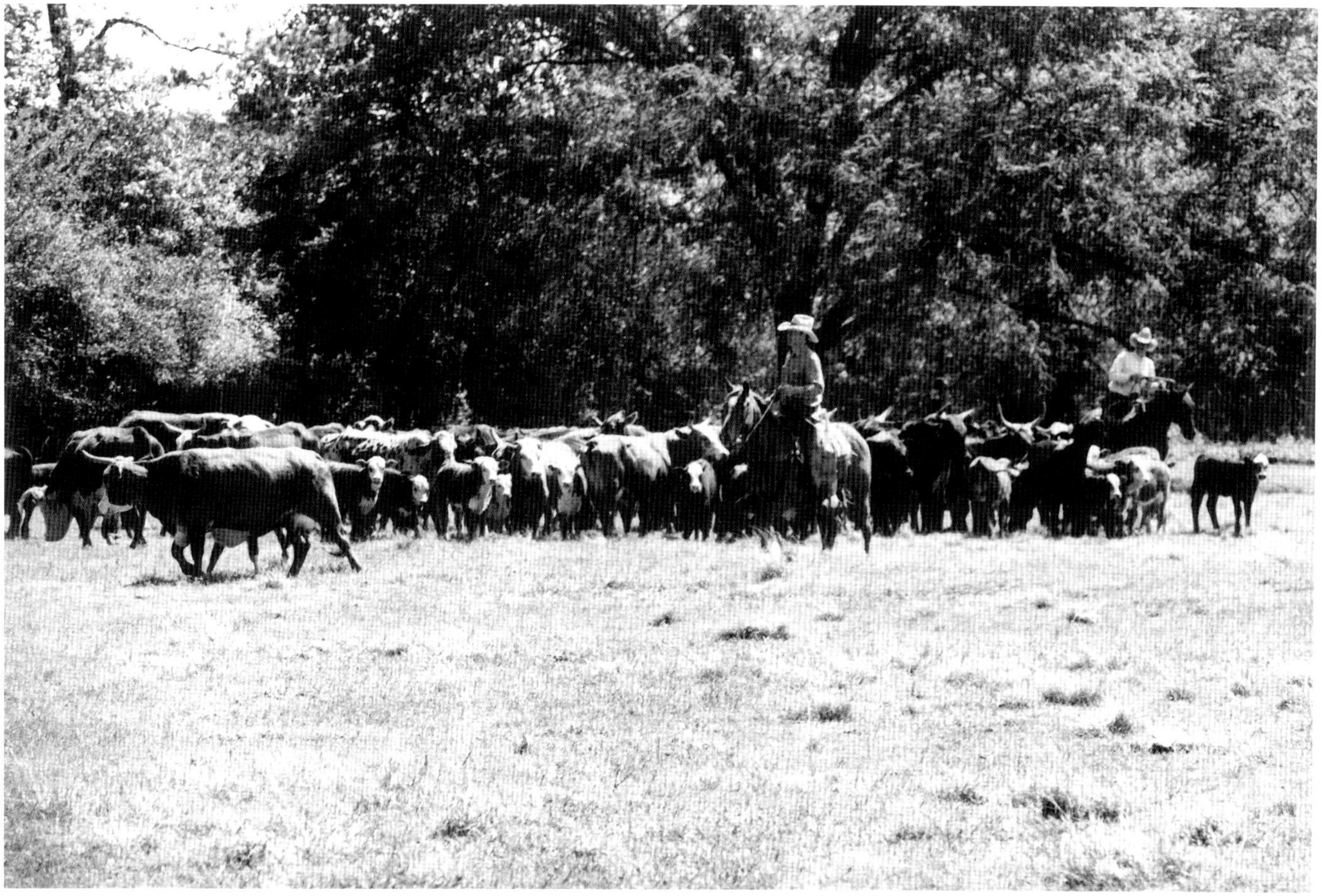

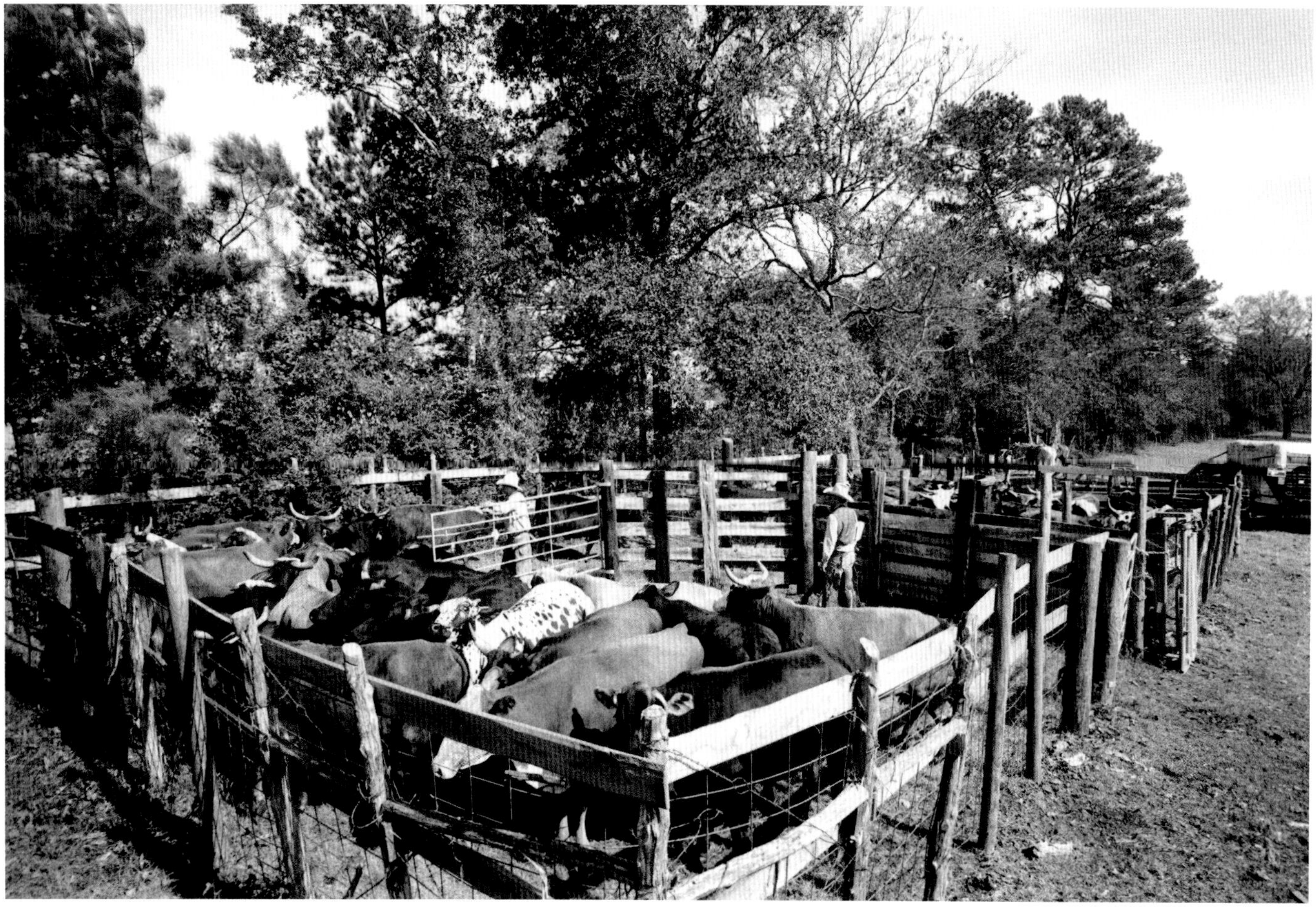

Needing a place for Guy Porter, they moved one of Hoyt's rent houses from Crockett to the ranch. It had set just off Camp Street behind Nook's Cafe, where the proprietress, Mrs. Nook, famously fed everyone and anyone, even the ones who couldn't afford the quarter for her equally famous sandwich of sausage, potato salad, and barbecue sauce, all arranged — miraculously, it must be assumed — between two pieces of white bread.

The brothers set to work clearing brush, fixing fence, rebuilding the pens. They went back to working cattle, falling back on what they'd been taught by Hoyt and Dayton Owens and Claude Searcy and Guy Pipes. Working horseback, working with their Blackmouth Curs, keeping the beeves bunched up. When they could, they kept playing music.

Guy Gillette was there, too, documenting the new works, the renewed place, documenting the family returned.

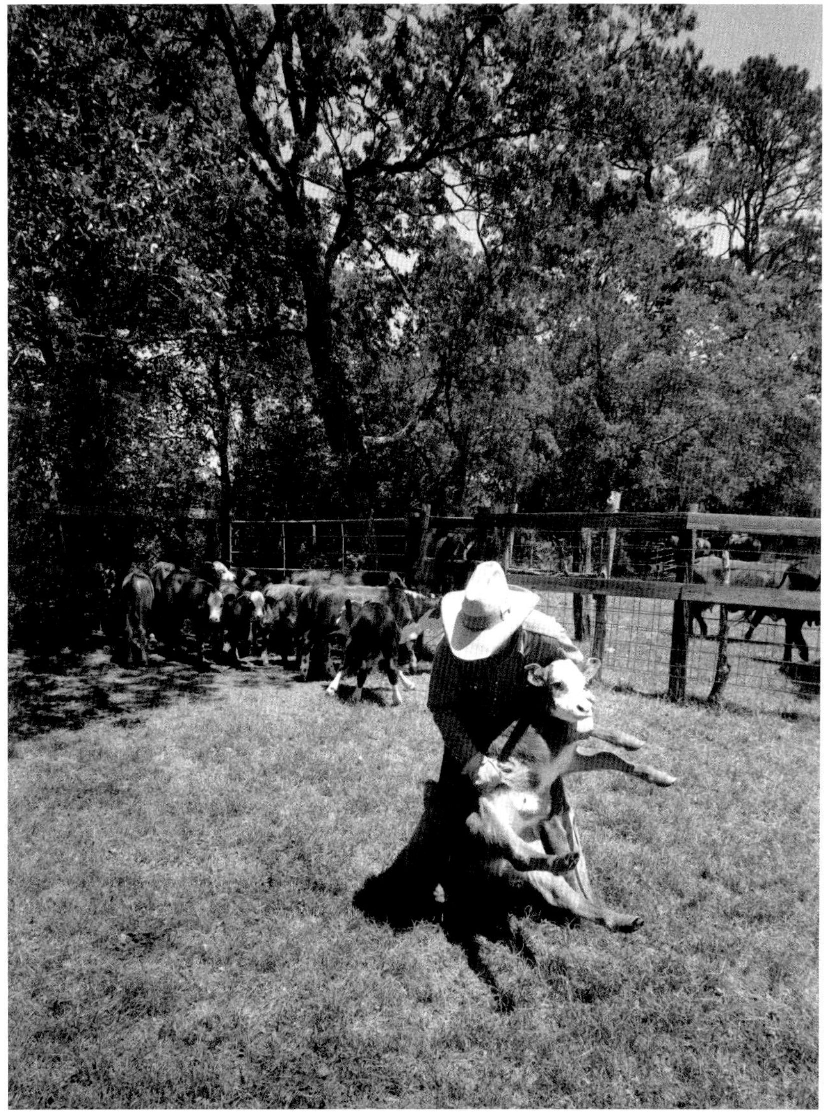

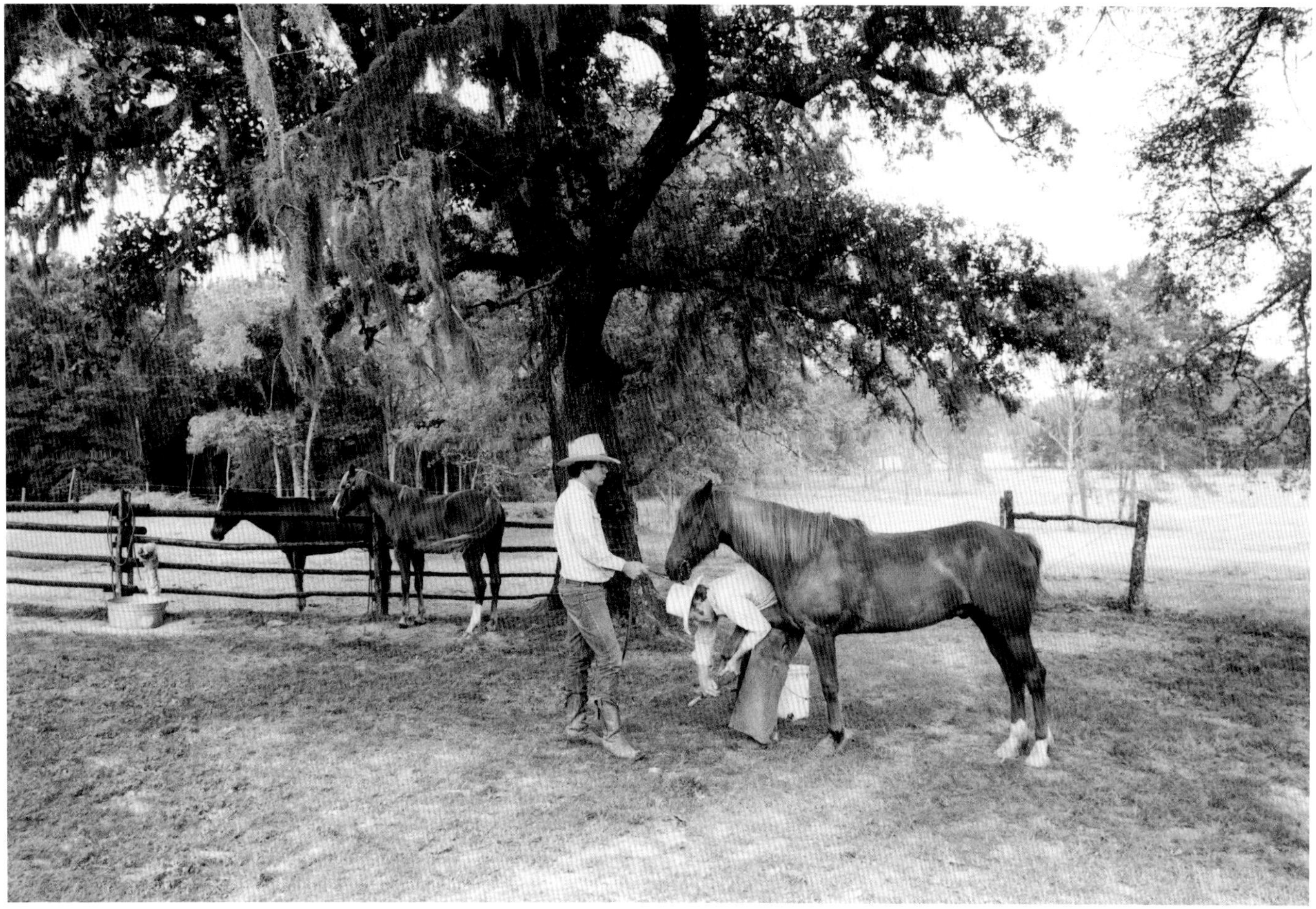

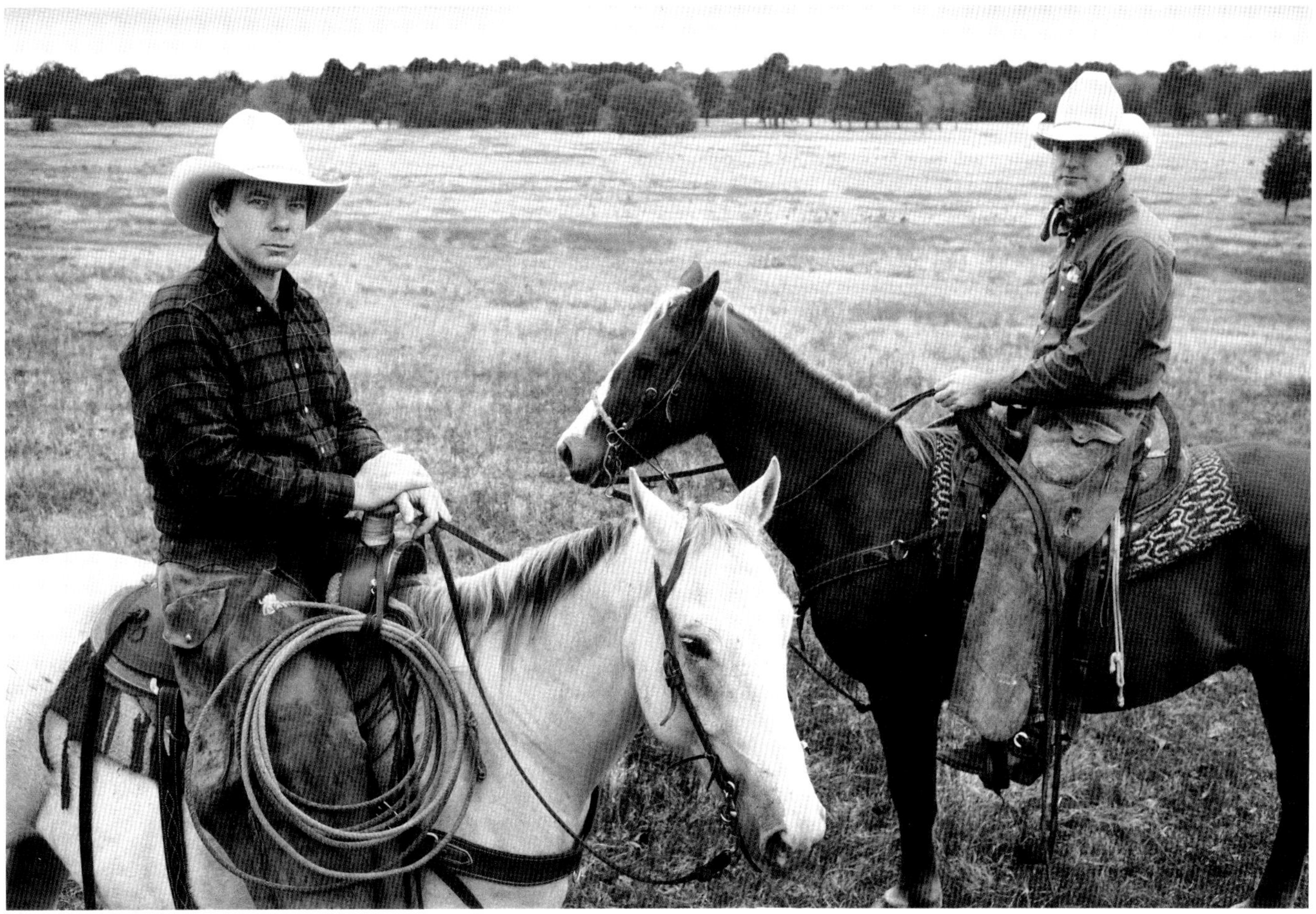

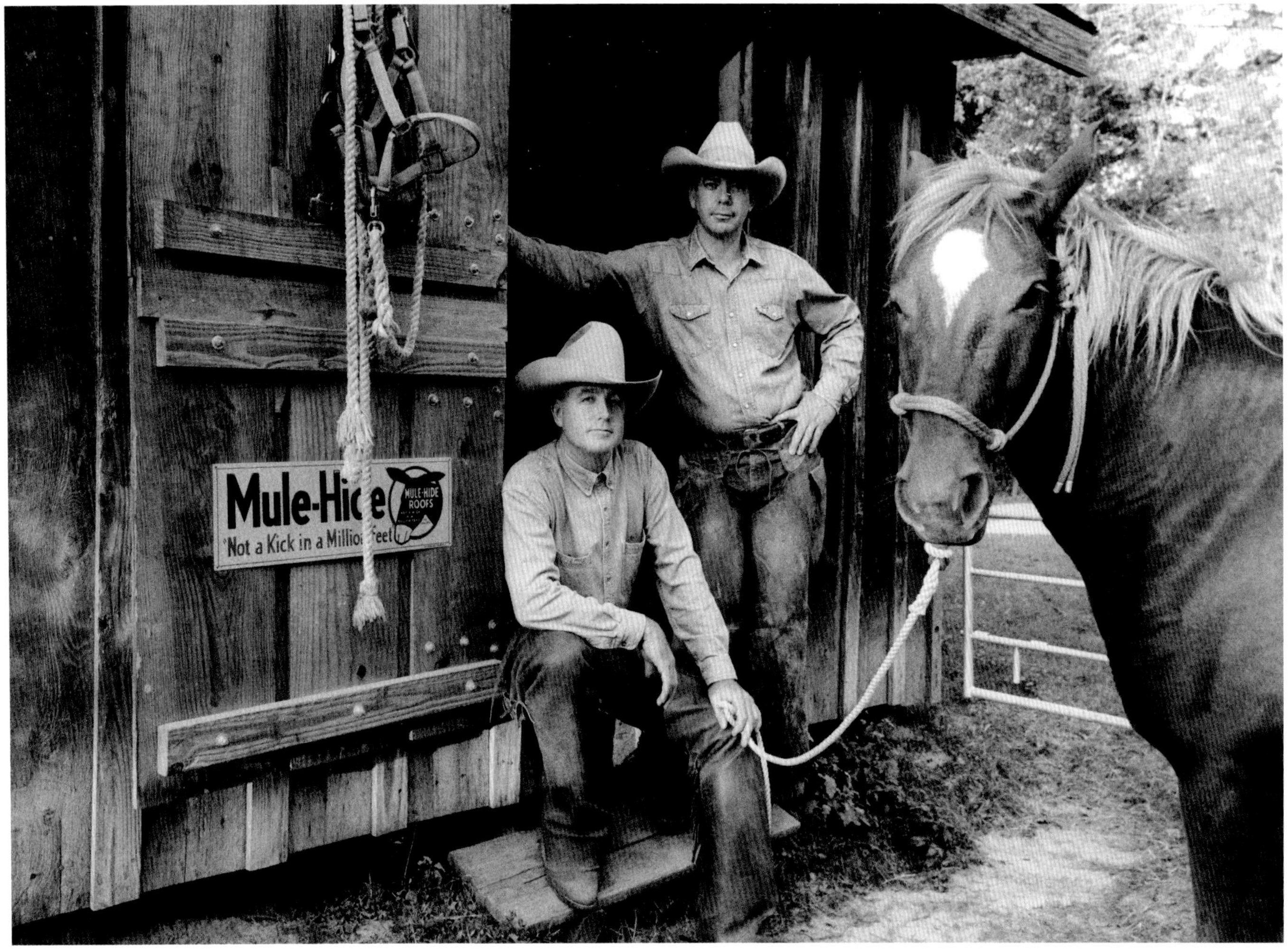
Mule-Hide
MULE-HIDE ROOFS
"Not a Kick in a Million Feet"

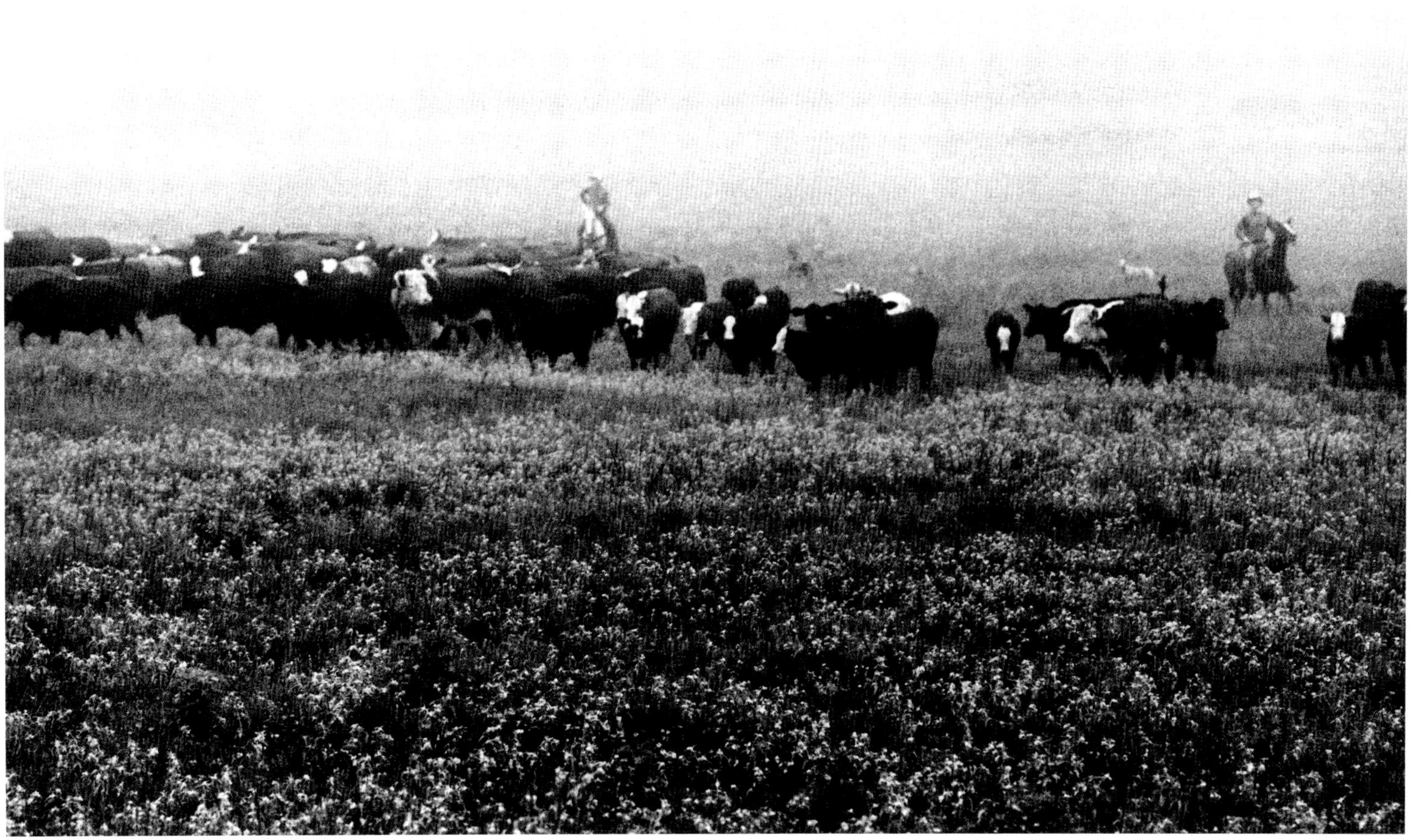

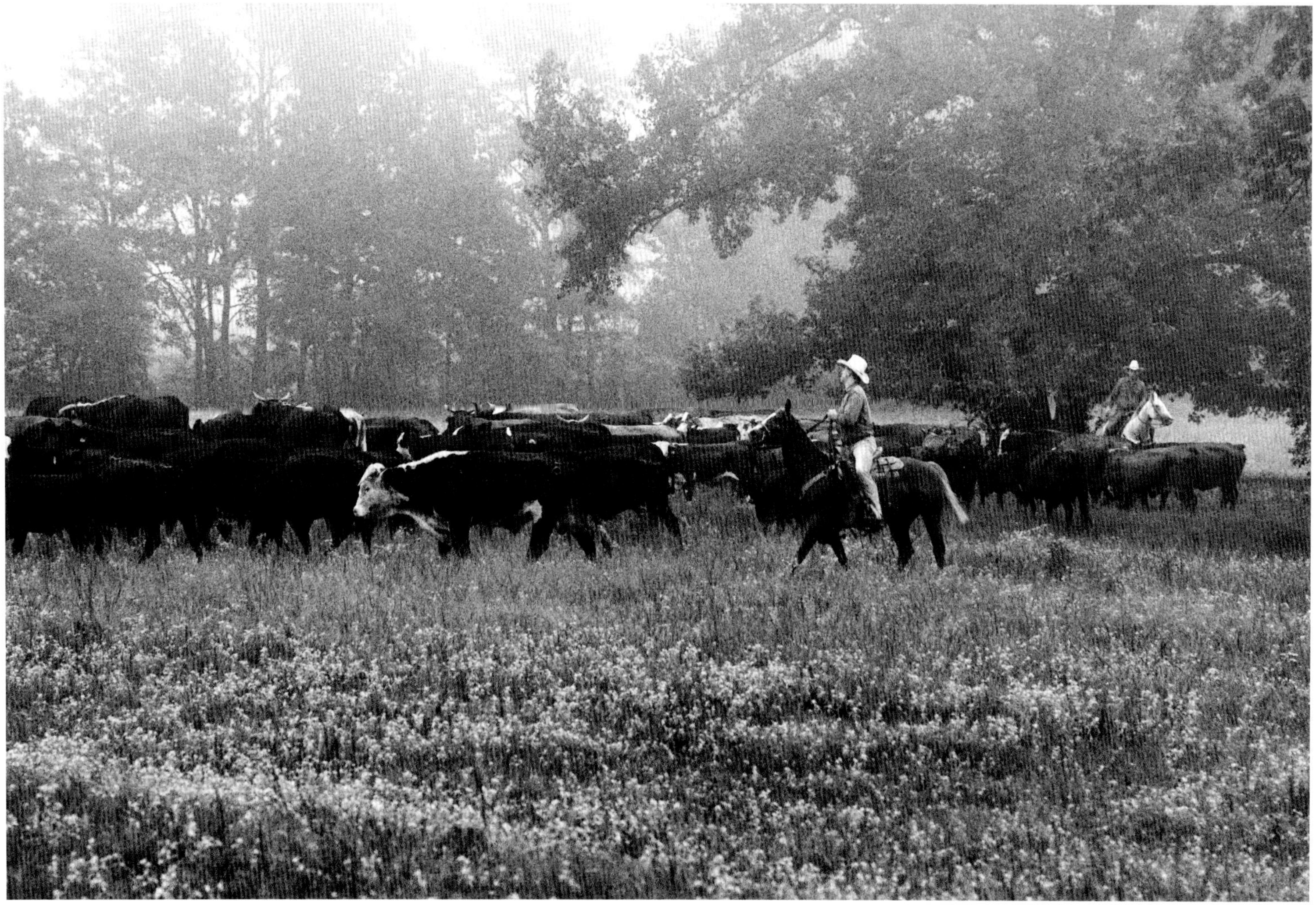

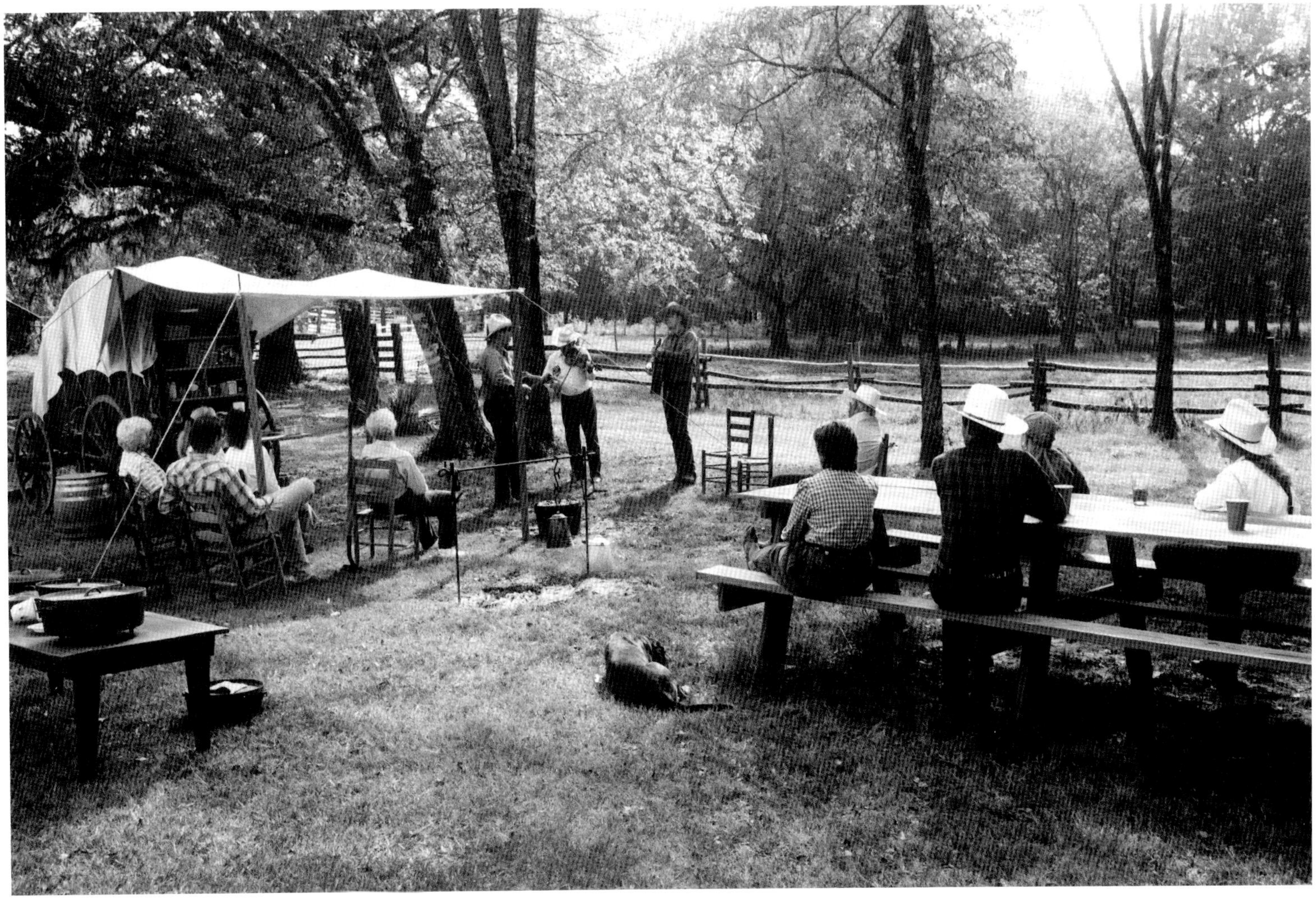

Next Generation

The Gillette Brothers still traveled some to do their music, but the land and the work of the land made it convenient to stay local when they could. Before opening their own place several years later in Crockett — the Camp Street Cafe — they often performed at a storied little joint in nearby Trinity. Called the Paper Moon when their mother, Doris, was a teenager, in the spillover of the Austin cosmic cowboy craze it got renamed the Armadillo Club. Proof positive that not all change is good. But good things happened there. The music was one. And one particular weekend some friends from Houston brought a young friend up for a visit to the ranch and a night at the club. That was another good thing.

Cathi was a Texas Panhandle girl, born in Shamrock and raised in Wheeler, though now working in the concrete canyons of Houston. She and Guy Porter began seeing one another and married a couple of years later. The ceremony was held at the Porter Place on a hill in the pasture where Guy Gillette, some thirty years before, had photographed two young boys learning from their grandfather how to cook breakfast on the range.

It was a cowboy wedding of the first water. The groom rode horseback to the outdoor altar; the best man drove the team that pulled the wagon carrying the wedding party. Once the vows were exchanged, groom and best man traded saddle and wagon seat. Twenty years later, the pretty flower girl, Cathi's niece Jessica, married the best man.

Guy Porter and Cathi named their baby girl Dorcie, Guy Gillette's nickname for Doris.

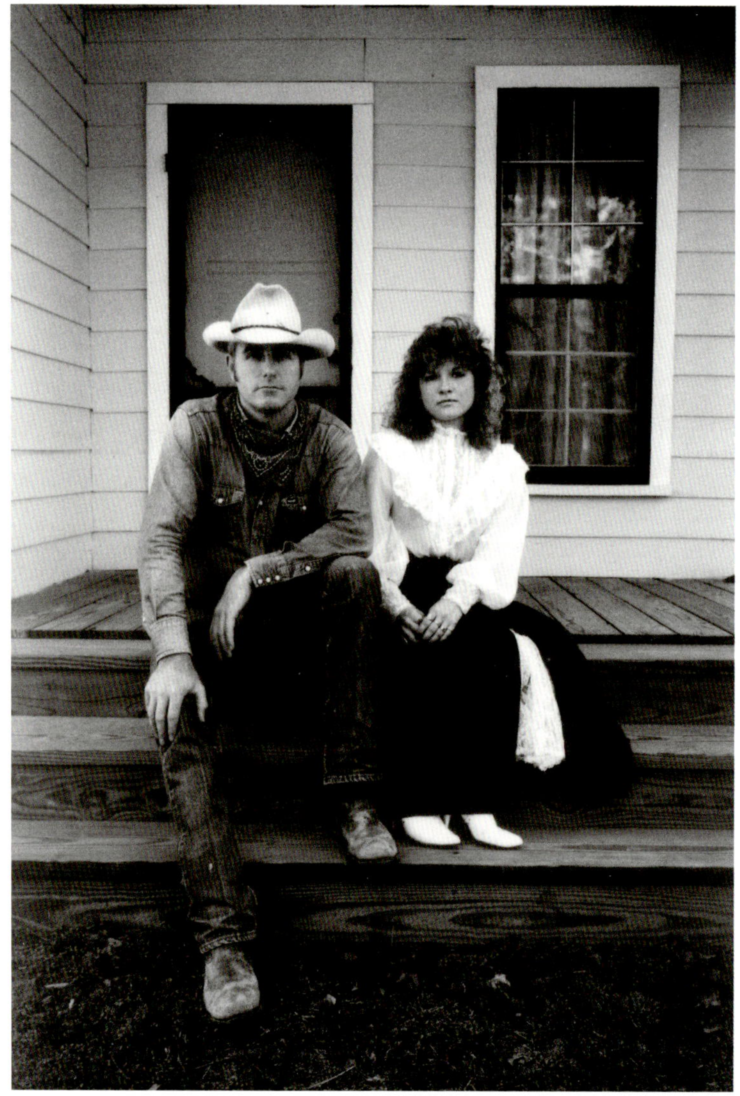

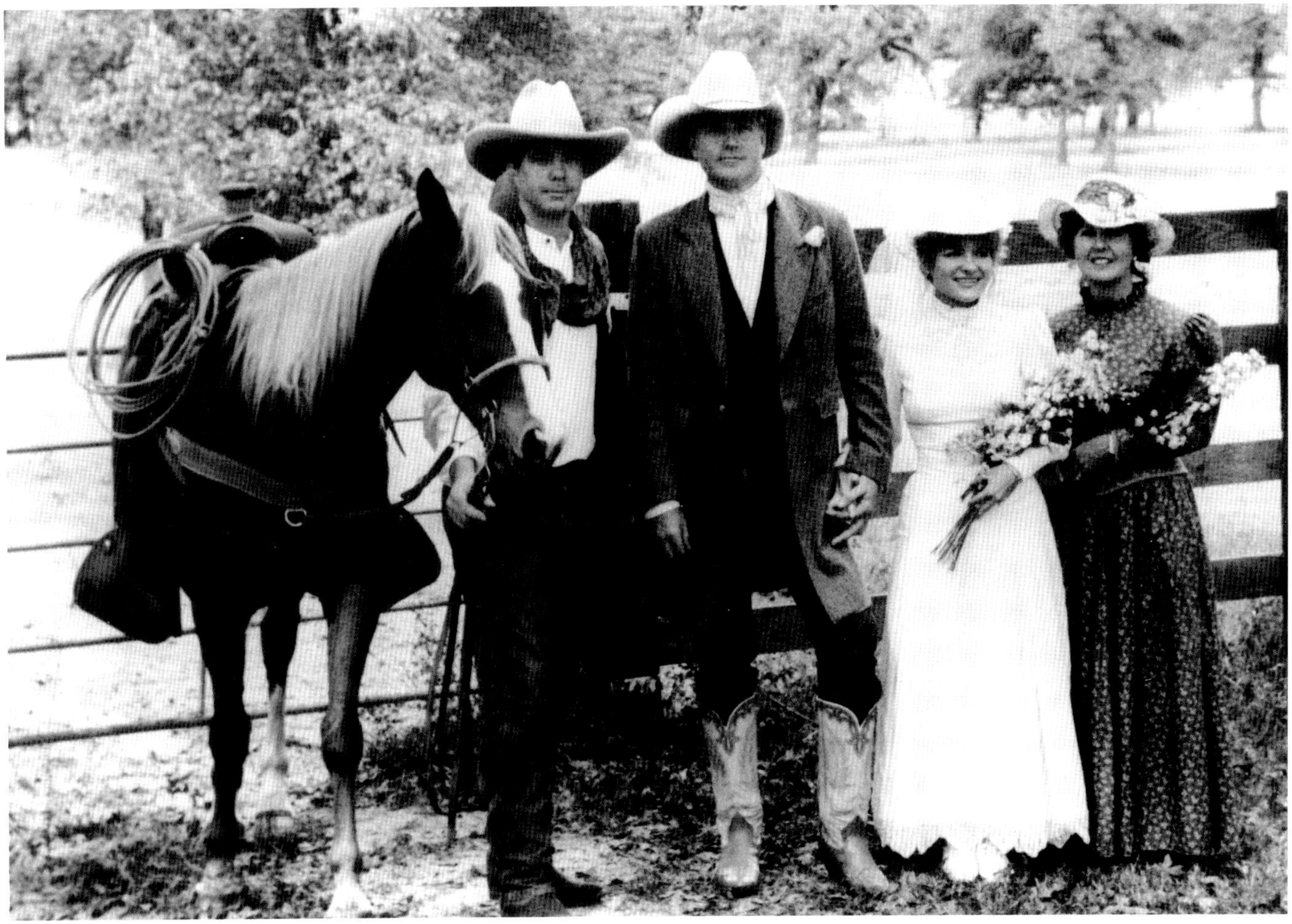

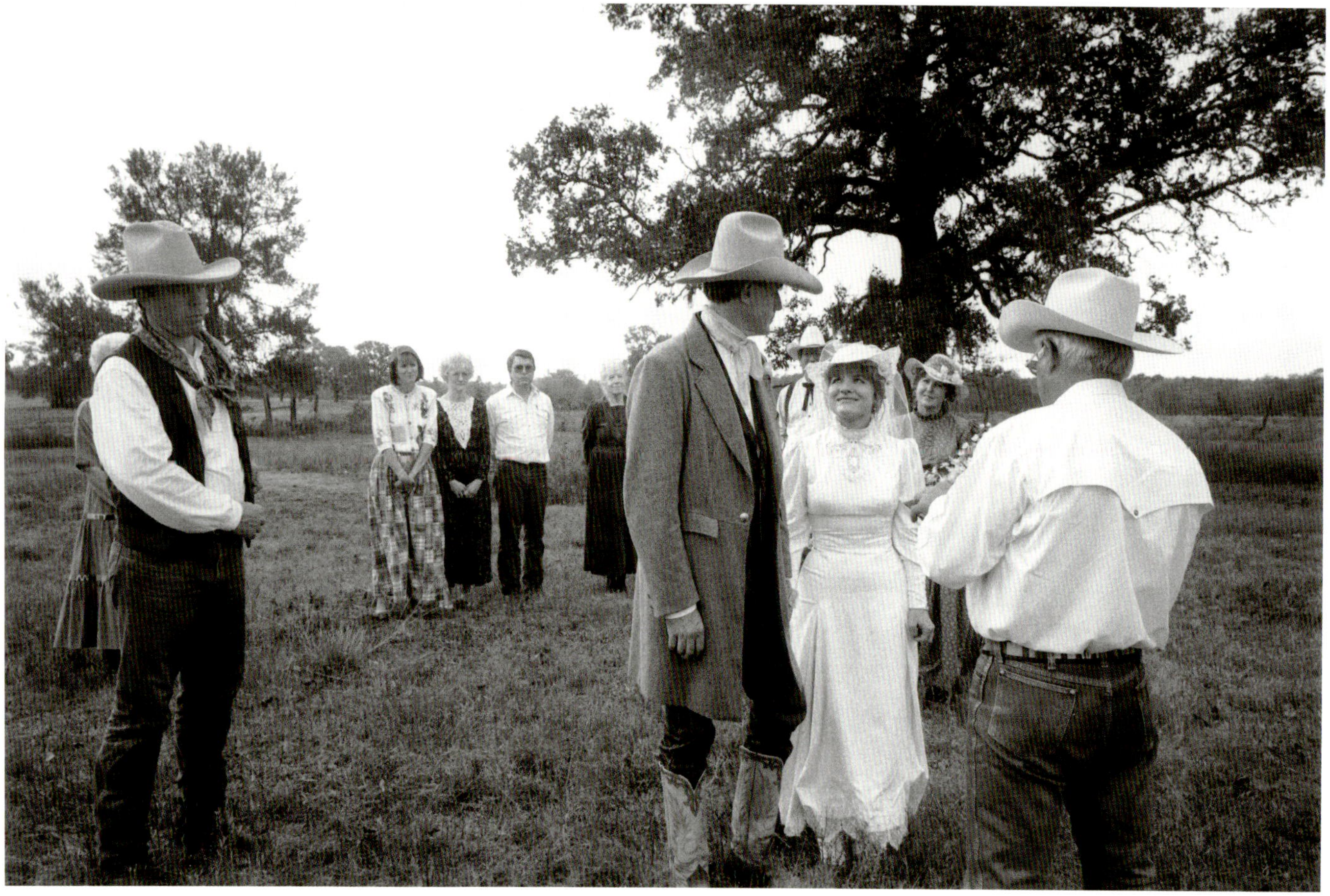

The universe is made of stories, not atoms.

MURIEL RUKEYSER

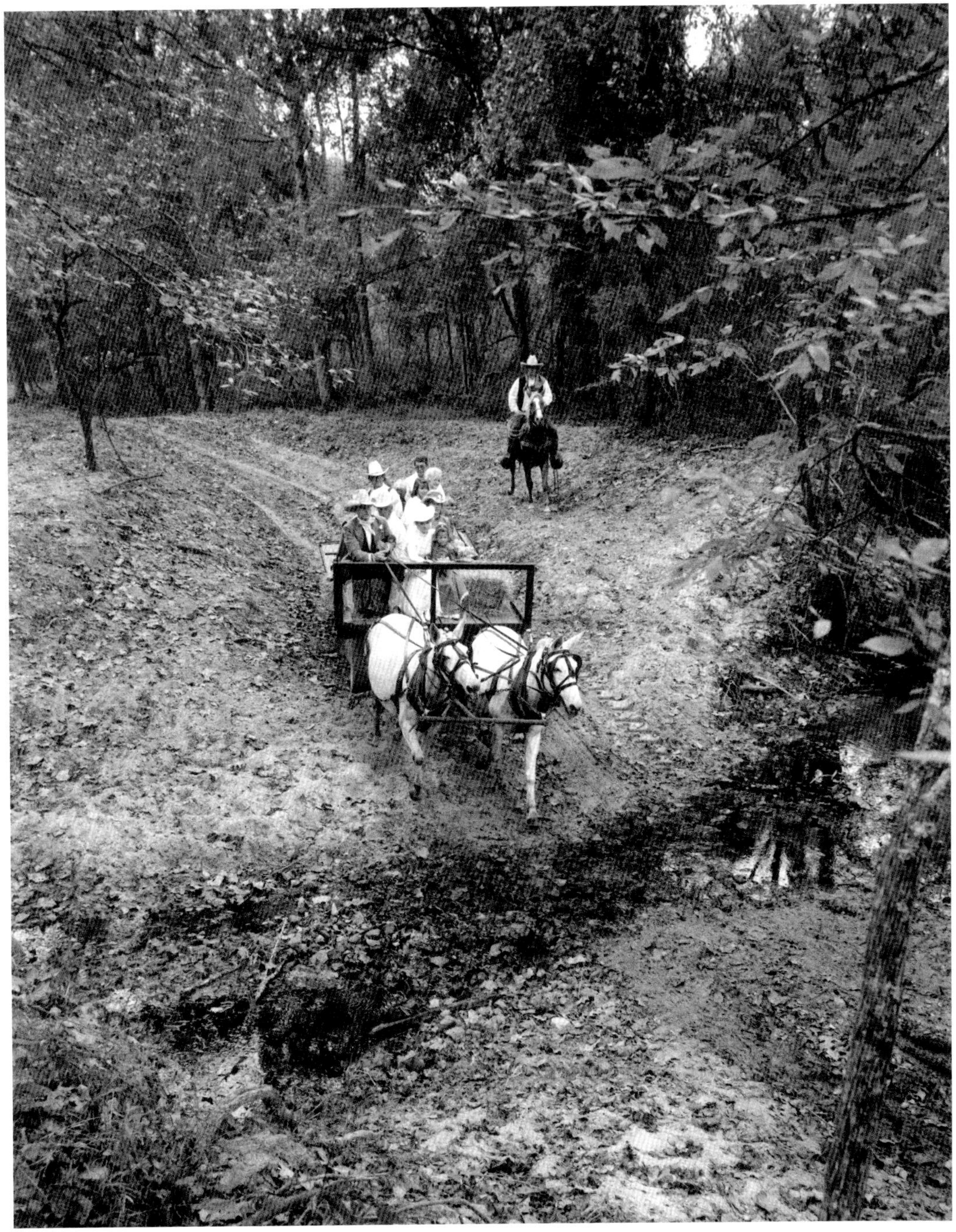

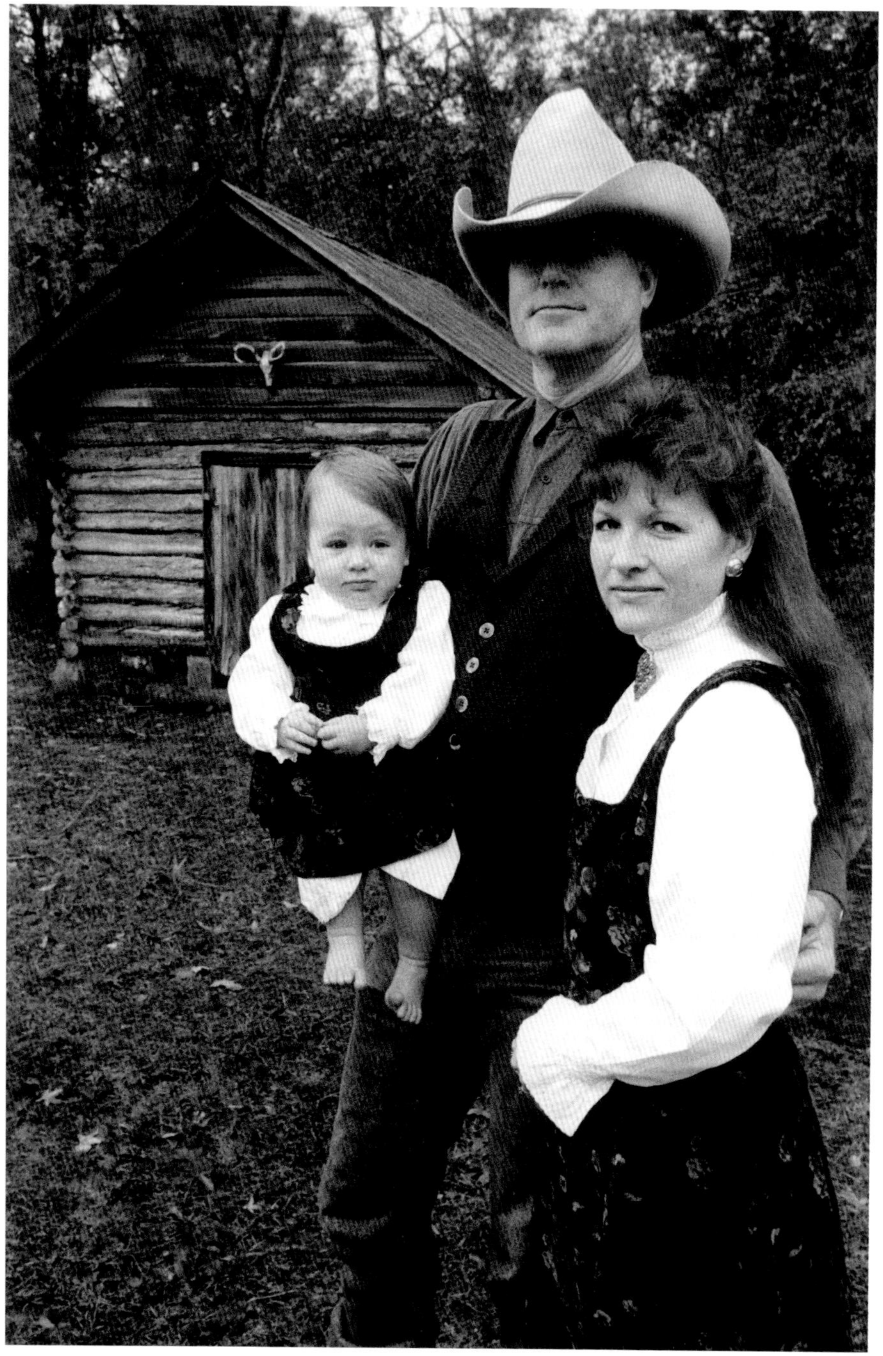

Eyes for the Blind

Doris, a city girl now for forty-plus years, sits in a rocking chair on the porch of the house where her son Pipp will live, where Guy Pipes lived before him, and Claude Searcy before Guy Pipes. She's in her working clothes, has been here a couple of months helping the boys make the Porter Place habitable again. She wears the slightest of smiles, so slight it's not really a smile at all, just a pleasant hint at the corners of the mouth. Enigmatic, you could say, or bemused. Perhaps, you'd think, because she's holding a pipe, a Meerschaum from the look of it, but not holding it like anyone who ever smoked a pipe would hold it. In her hands, it's ready for work, held like a screwdriver or a paint scraper. The rocking chair is awkwardly set behind the corner post, correct for the photograph but not how you'd angle it to pass the cool of the late afternoon as the sun starts down on the other side of the house. The cow skull, too, is too well placed. Props. The hand, you could say, of the photographer. Photographing the re-inhabiting.

Or, you could say, the photographer's eye. What makes the camera work. Better still, you'd think, Guy Gillette's eye, the eye of an actor-turned-photographer who sees the scene without having to know the script, whose long apprenticeship has been to the story, to the telling of the story in composition and light and tone and line, to the story told in the making of good pictures.

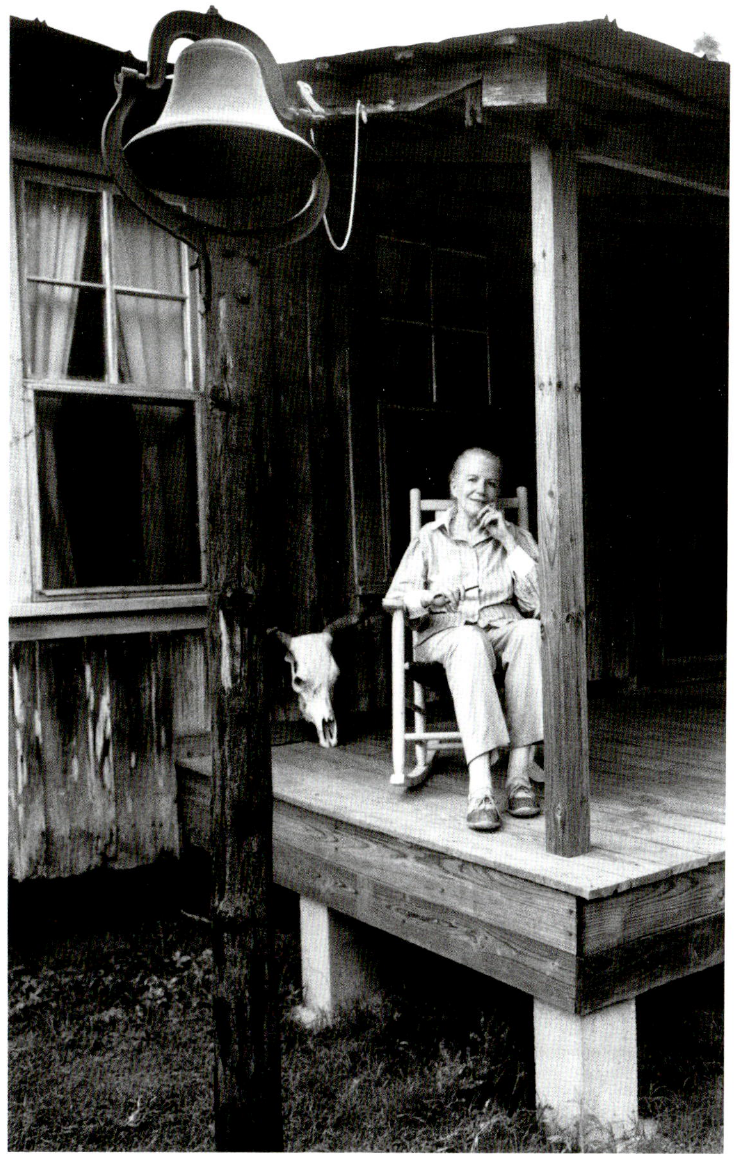

Like these other pictures. Hoyt and a young Pipp surveying the pasture, surveying the land, surveying, you could say, the future landscape. A future landscape that comes of two boys growing up on the land, guided by the hands of those who love them, who love the land. A future landscape that comes of learning their gear, their work. A future landscape that comes of boys growing into young men — a portrait of a teenaged Guy Porter who punched cows for wages for Watt Matthews at Lambshead, a portrait of a full-grown Pipp by the working pens. Each photograph a note in a melody, each uniting what has come before with what will come after. In the coda, on the porch on the back of Guy Porter's house, looking north past the barn and the corral and the smokehouse, the family re-inhabiting the land. At least for a little while.

Now, look back to the first photograph in this last series. To what you can't see in it, but to what you can feel. Catty-corner from Doris's left shoulder, out of the picture though only a few dozen yards away, is a house of similar design and age, a rambling box-and-strip with a porch wrapping two sides, the south and the

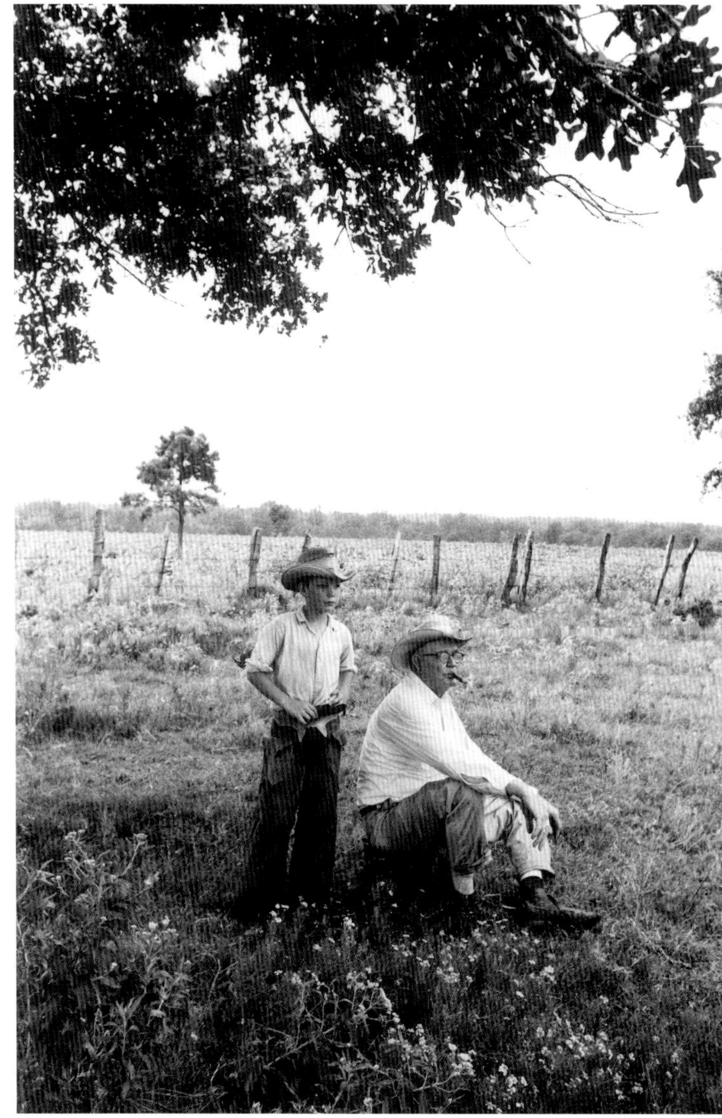

east, screen doors all around. It once sat atop the hill in the pasture where Guy Porter and Cathi later exchanged wedding vows. In the 1950s, Hoyt moved it here for practical reasons, so that more than one house could be near the barn, so that both could draw from the same well, so that another house could serve as free residential quarters for another hired hand. Julian Lewis and his family lived here for a decade. A black sharecropper who worked Porter Place land, Julian moved away when the sharecropping days ended, relocating up the country in Grapeland. Pipp invited him down for a visit in 2004, the first time Julian Lewis had been back, the first time he'd seen the place in thirty years. "Eyes for the blind!" he said, "Eyes for the blind!" You could say that, too. Eyes for the blind.

You are in the photographic business and a master of it.

J. FRANK DOBIE,
in a letter to Guy Gillette, May 30, 1963

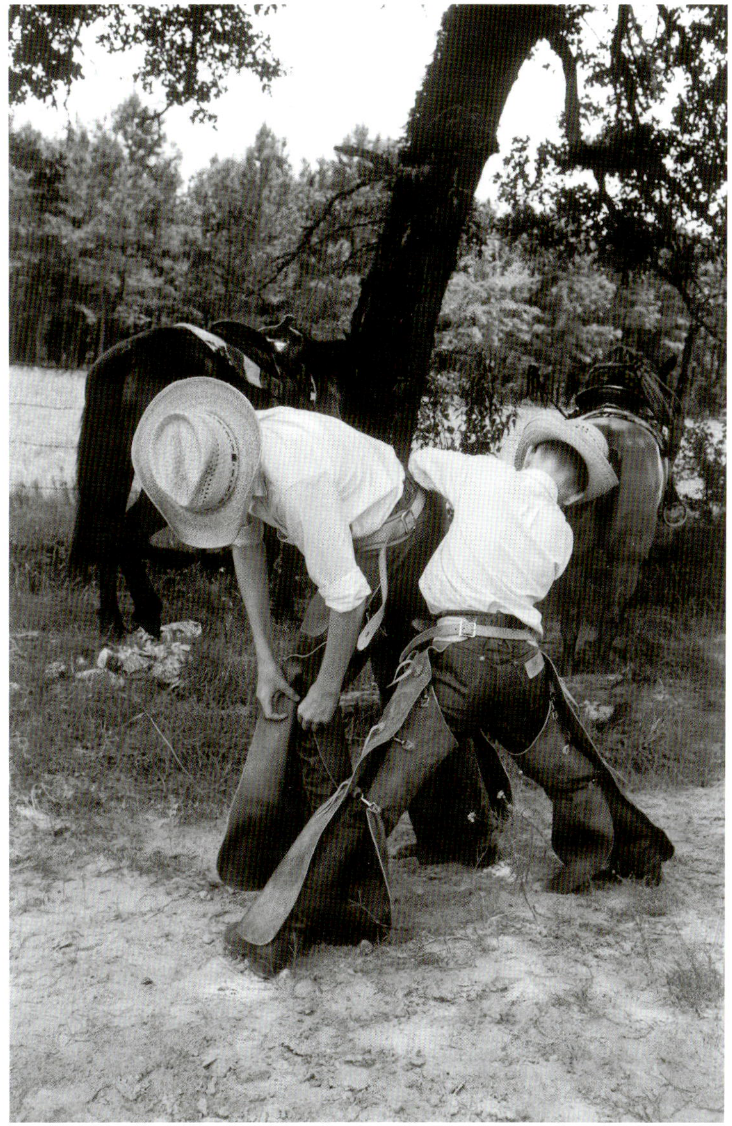

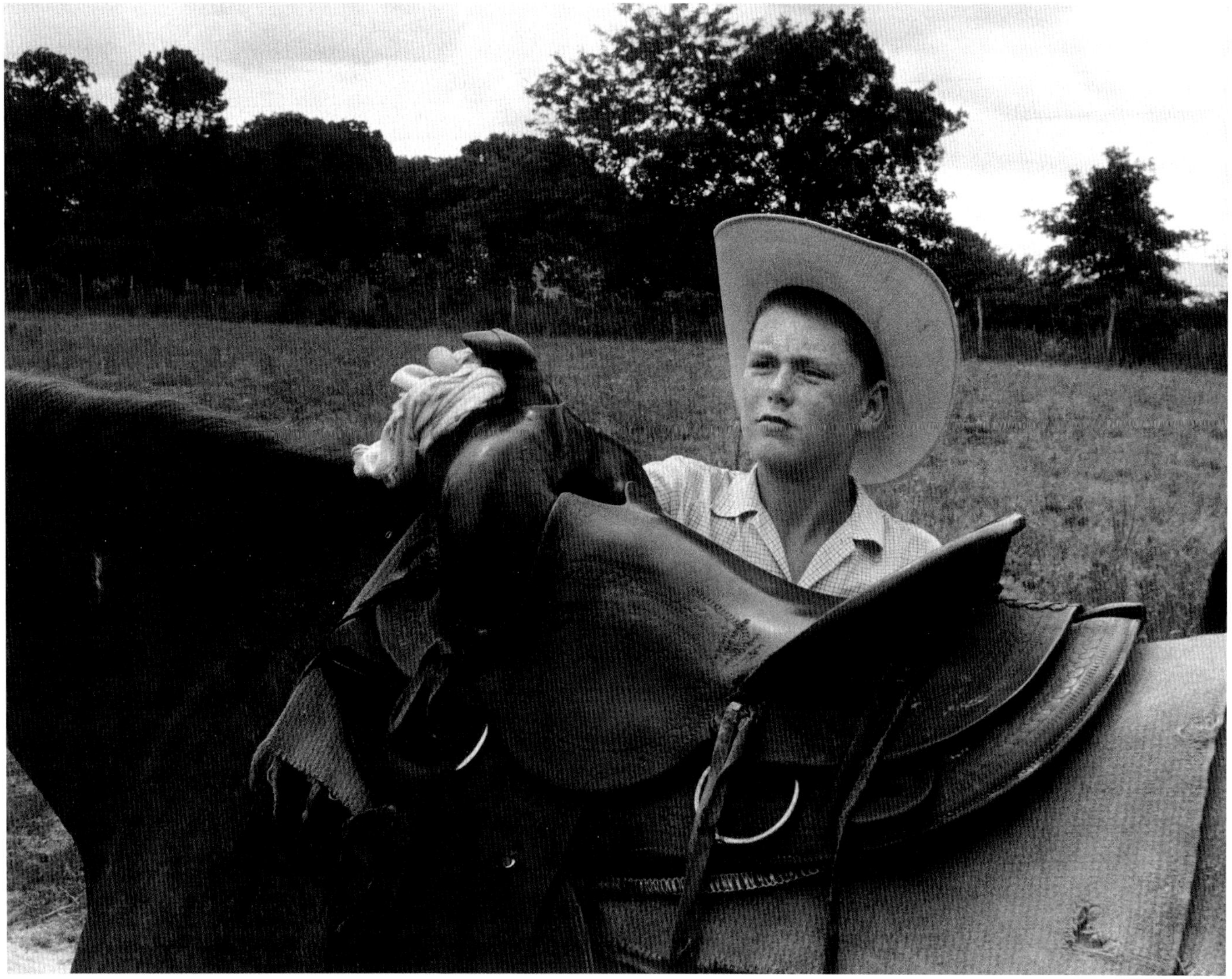

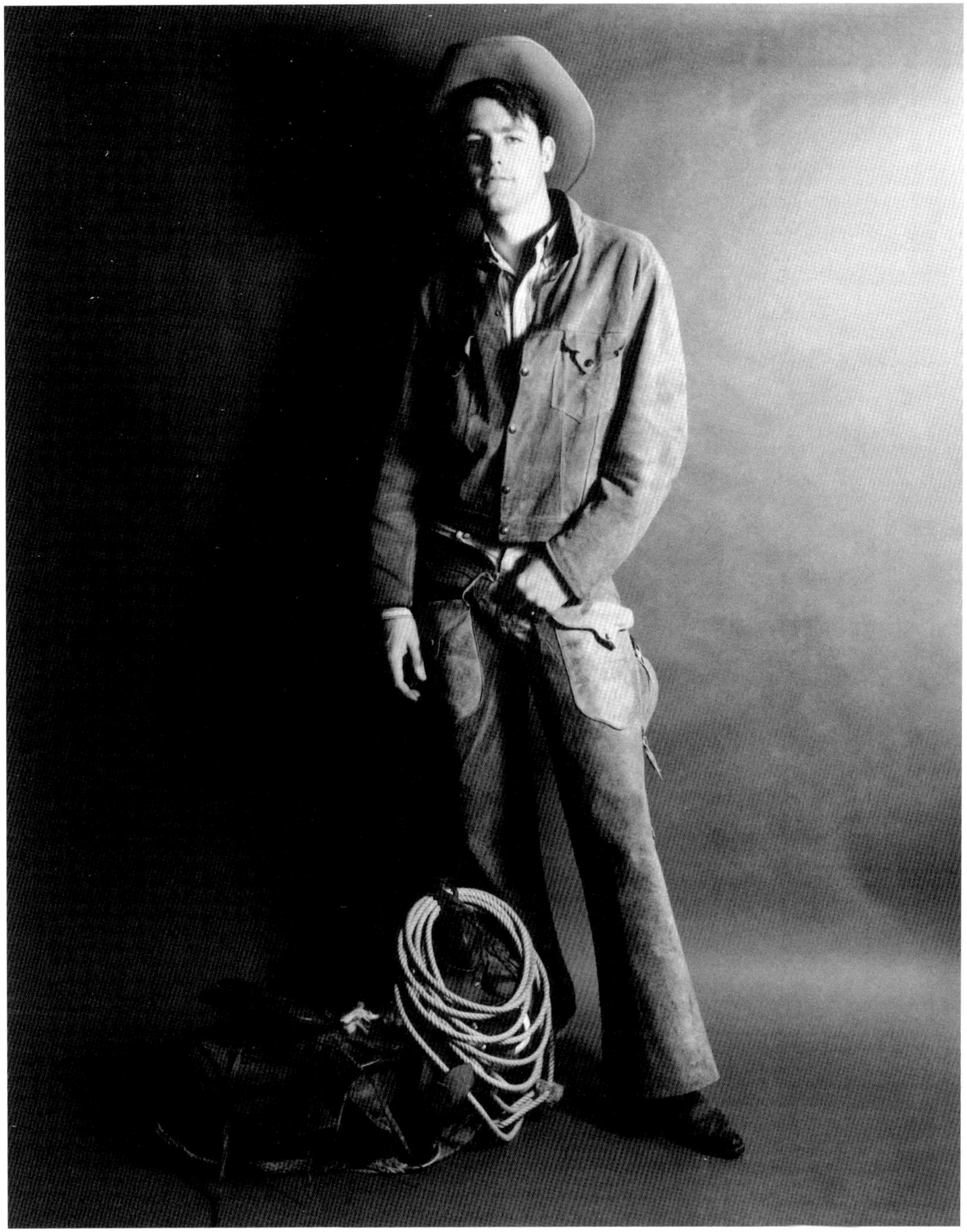

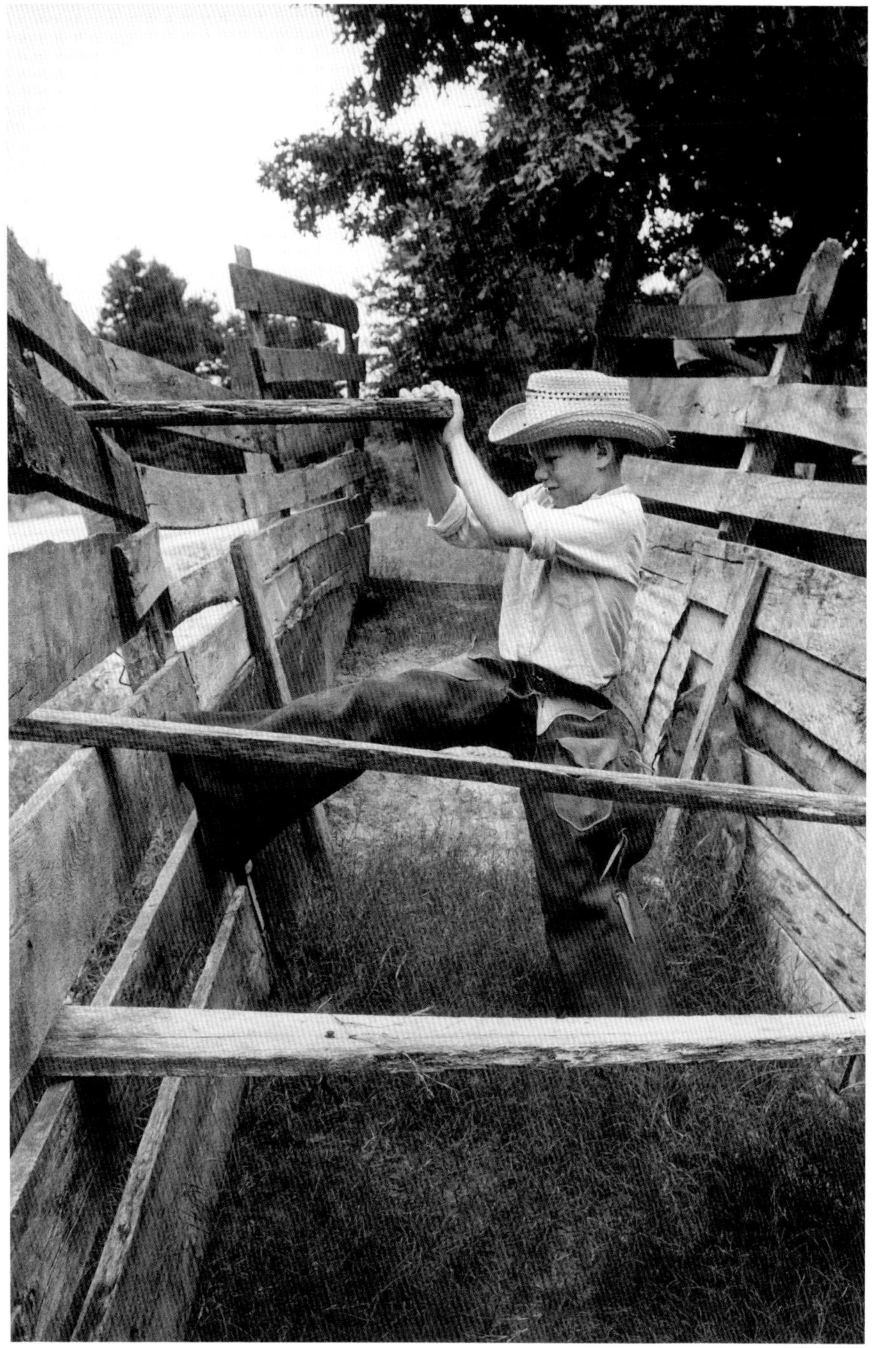

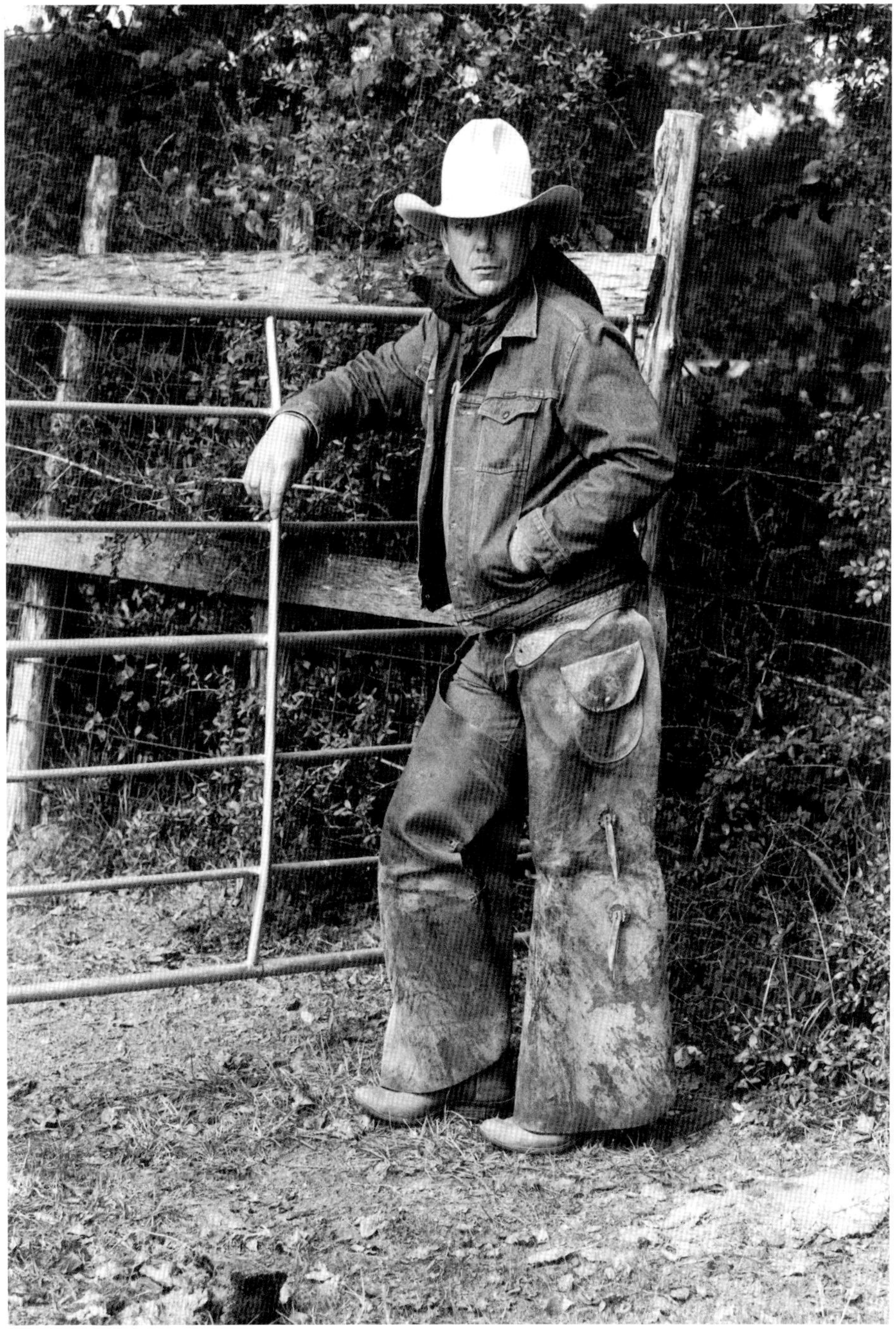

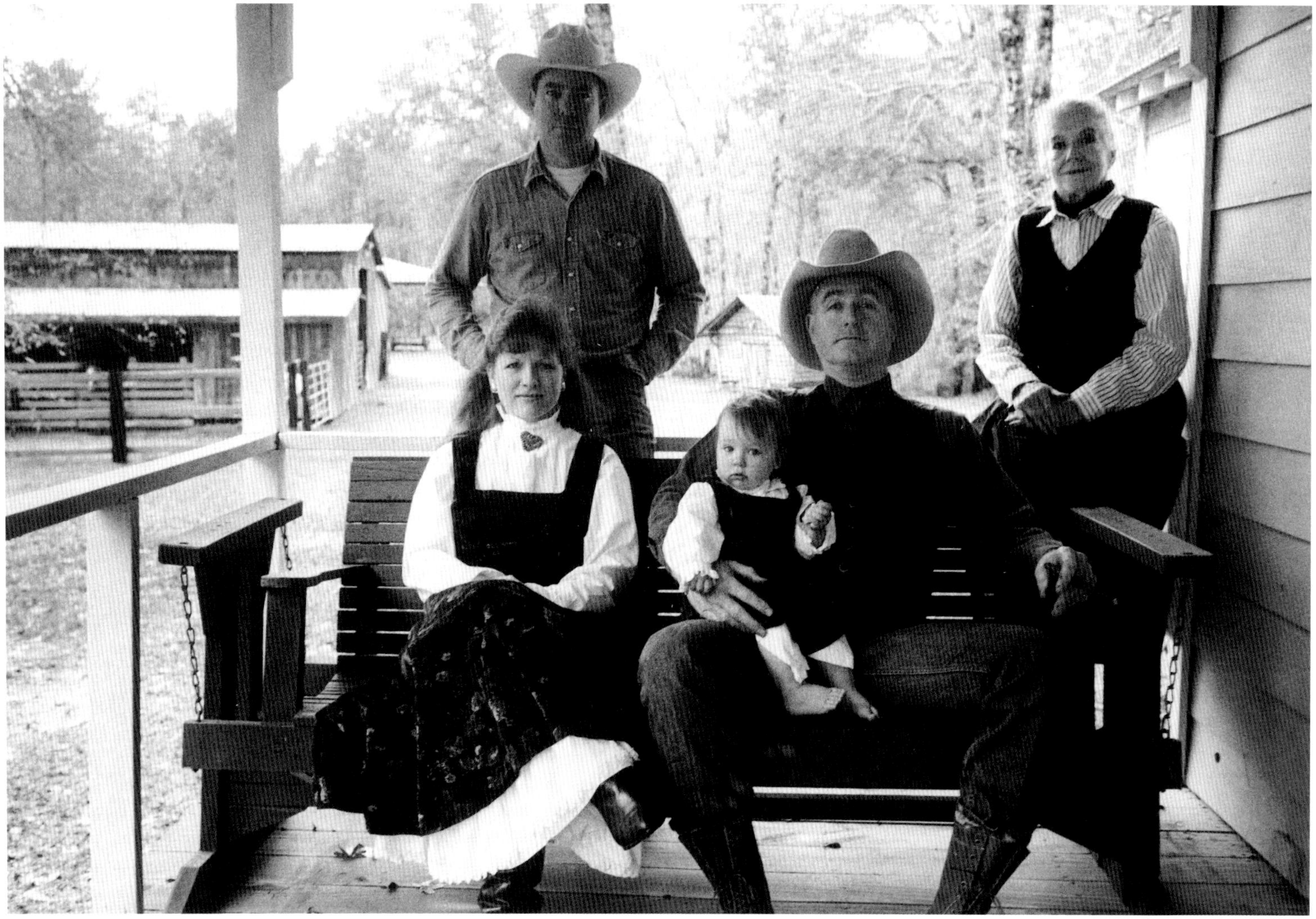

What Still Remains
AN AFTERWORD

In late August of 2012, Doris Porter Gillette died after a short illness. She was ninety-one, married seventy of those years to Guy Gillette. As summer ended and fall began, Guy made the long journey from Yonkers, New York, to Texas, to Crockett, to spend some time with his two sons and their wives and his granddaughter. For another visit to the Porter Place.

He arrived during the stumbling recovery from the drought of 2011, the worst since the 1950s, and during a year still reeling from the deepest financial crisis since the 1930s. As we go to press, it's still tough on the Porter Place, just like it is on every ranch across North America. Ranching is never easy. Still, it's as tough now as it's been in a long, long time.

Hoyt and Lucy Porter, who built this place, are gone. So are others who had their hand in it: Dayton Owens and Claude Searcy and Guy Pipes. And now Guy's beloved Dorcie is gone, too.

Still, people want beef. And, someday, it will rain again. In the stories that Guy Gillette has left us in these photographs, we find hope for that day, the good day that is to come. For he leaves us not with nostalgia for what was, but with the reminder of what still remains: the land, the labor of the land, and this family of the land.

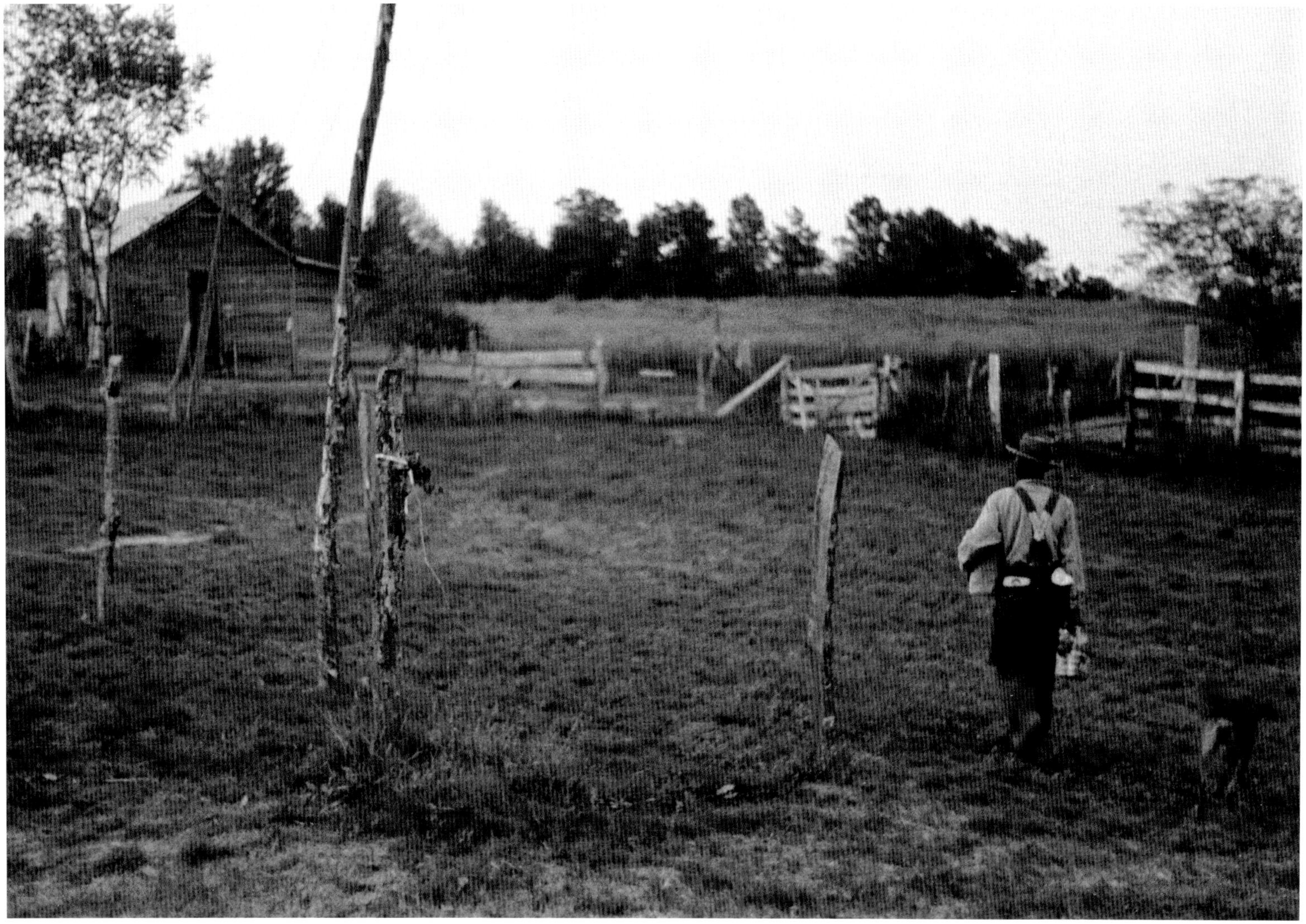

About the Photographs
AN APPENDIX

Telling a story was always
the attraction of photography for me.

GUY GILLETTE

COPYEDITED BY *Pippa Letsky*

TEXT DESIGN AND COMPOSITION BY *Julie Rushing*
Set in Kepler and Douglass Pen

JACKET DESIGN BY *Tony Roberts*

IMAGE PREPRESS BY *University of Oklahoma Printing Services*

PRINTED AND BOUND BY *Tien Wah Press (PTE) Limited, Singapore*

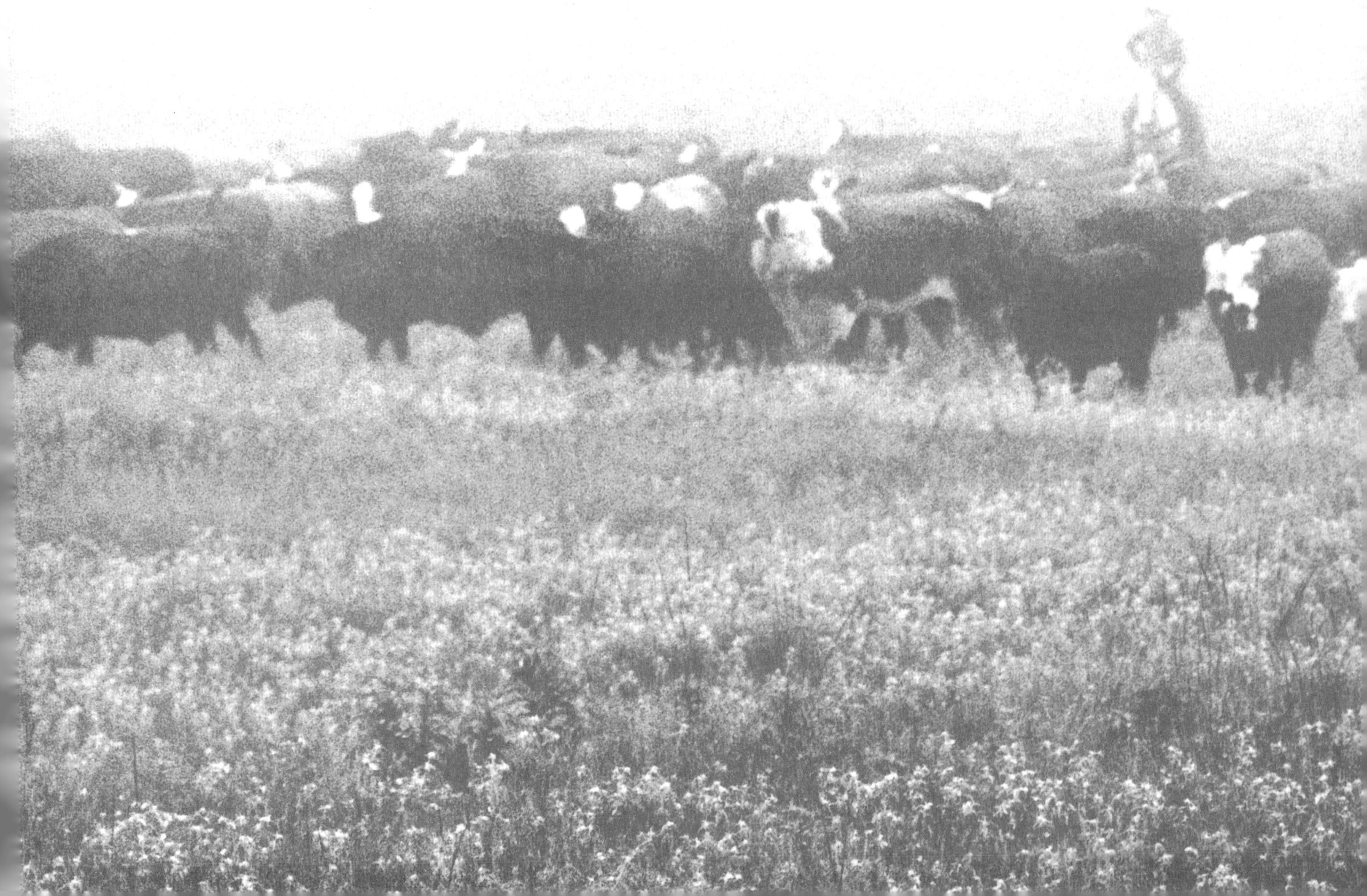